ESSENTIAL

OPPORTUNITY CLASS

MATHEMATICAL REASONING

5 Mathematical Reasoning Test Papers

RAY LEE | JIMMY LIU

To Cookie, my light,
and to Phoebe who I hope
can one day figure out all the questions in this book.

Five Senses Education Pty Ltd
2/195 Prospect Highway
Seven Hills 2147
New South Wales Australia

First Published 2023

Lee, Ray and Liu, Jimmy

Essential Opportunity Class
Mathematical Reasoning Book 2
ISBN 978-1-76032-529-9

2024 03 25

Contents

Preface

This book is designed to help students prepare for the Opportunity Class Placement Test. It consists of five mathematical reasoning practice exam papers and is suitable for use by Year 3 and 4 students. The exam papers are designed to the exact format of the Opportunity Class Placement Test, with hand picked questions that closely relate to past Opportunity Class Placement Test questions.

Success in this extremely competitive exam requires commitment and hard work. We hope these practice exam papers can help you achieve your goals.

Five Senses Education

OC Practice Test

Mathematical Reasoning 6 (Time allowed: 40 min)

INSTRUCTIONS

1 Write your Name on the cover page.

2 There are 35 questions in this paper. For each question there are five possible answers, A, B, C, D and E. Choose the one correct answer and record your choice on the separate answer sheet. If you make a mistake, erase thoroughly and try again.

3 You will not lose marks for incorrect answers, so you should attempt all 35 questions

4 You must complete the answer sheet within the time limit. There will not be any extra time at the end of the exam to record your answers on the answer sheet.

5 You can use the question paper for working out, but no extra paper is allowed.

Name: ______________________________

1 The factors of 6 are 1, 6, 2 and 3.

Which of these numbers has the most factors?

A 12

B 15

C 16

D 19

E 21

2 Harry had four tiles.

He joined the tiles together with no gaps or overlaps to make the following shape.

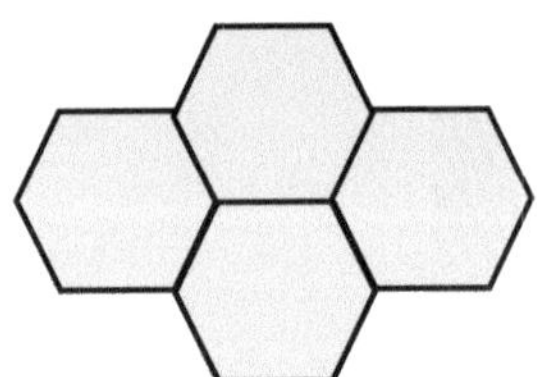

What is the number of edges Harry's new shape have?

A 12

B 14

C 16

D 18

E 20

3 Which number will come next in this sequence?

1	2	4	5	25	26	?

A 512
B 576
C 676
D 812
E 1002

4 What fraction is shaded?

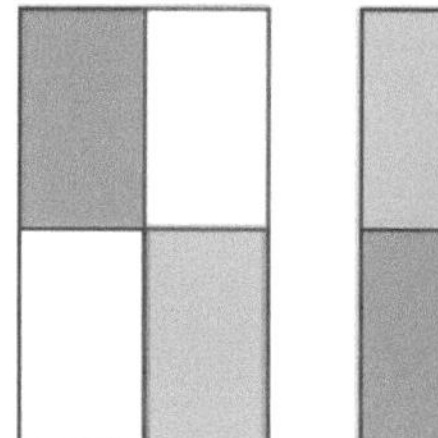
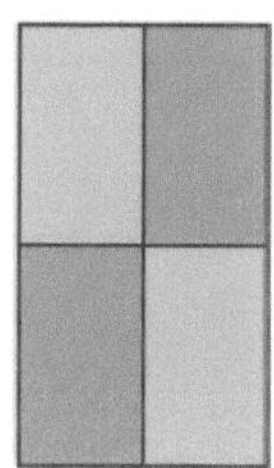
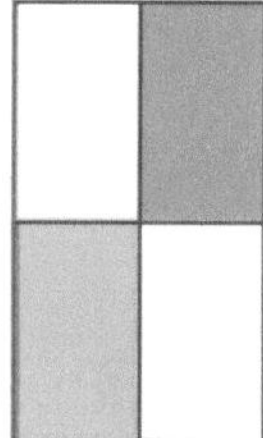
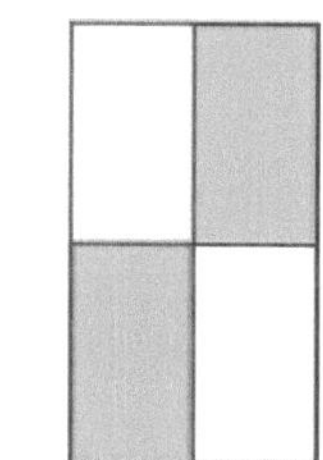

A $\frac{4}{8}$

B $\frac{12}{16}$

C $\frac{5}{6}$

D $\frac{3}{6}$

E $\frac{5}{8}$

5 Jessica is using a number code to represent letters. Each letter is shown by a number. This is how Jessica would code the word "TODAY":

T O D A Y
20 – 15 – 04 – 01 – 25

How should Jessica code the word "WATER"?

A 23 – 01 – 19 – 20 – 18

B 23 – 01 – 20 – 05 – 18

C 23 – 01 – 20 – 19 – 18

D 23 – 01 – 20 – 18 – 19

E 23 – 01 – 18 – 05 – 20

6 The picture shows some shapes.

Bill picks one of the shapes without looking. What is he most likely to pick?

A a circle

B a triangle

C a black shape

D a white shape

E a white triangle

7 When Jason started primary school he was 120 cm tall.

When Jason finished primary school he was 160 cm tall.

By how much has Jason's height increased?

A $\frac{3}{4}$ of 160 cm

B $\frac{1}{2}$ of 120 cm

C $\frac{1}{4}$ of 120 cm

D $\frac{2}{3}$ of 120 cm

E $\frac{1}{3}$ of 120 cm

8 Trisha started to make a square using a pattern of dark and white tiles.

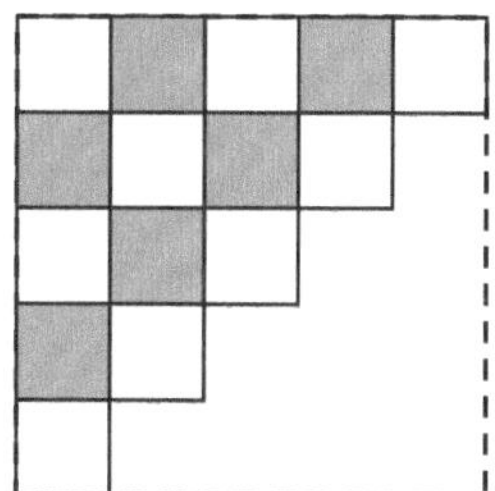

How many MORE white tiles does she need to complete the square?

A 6

B 5

C 4

D 3

E 2

*Use the information below to work out questions **9** to **11**.*

Steven is planning to build a house. His architect gives him the following costs when designing the house. He has $500 000 to spend and the house must include at least 3 bedrooms, 2 bathrooms a kitchen and 2 living spaces.

Room	Cost
1 Bedroom	$ 45 000
1 Bathroom	$ 60 000
1 Kitchen	$ 80 000
1 Living Space	$ 50 000

9 How much will the three bedrooms cost?

A $ 135 000
B $ 136 500
C $ 137 000
D $ 135 500
E $ 139 500

10 What is the minimum cost of Steven's desired house?

A $410 500
B $450 000
C $435 000
D $485 500
E $500 000

11 With the left over amount, Steven wants to build the most expensive room he can.

What can he build?

A 1 Bedroom
B 1 Bathroom
C 1 Living Space
D 1 Kitchen
E Nothing

12 What is the area of this shape in square metres?

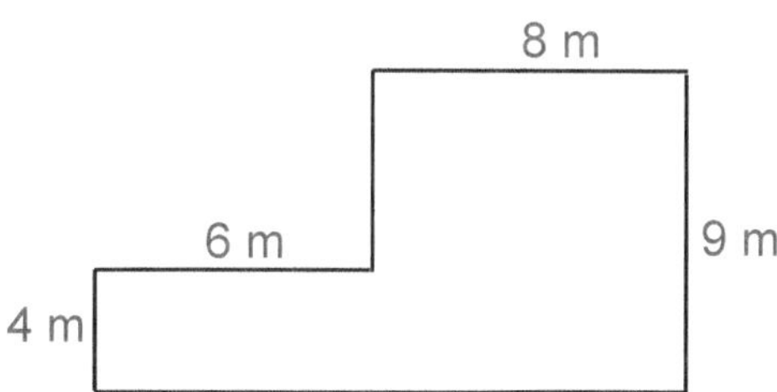

A 182 m^2

B 114 m^2

C 96 m^2

D 72 m^2

E 24 m^2

13 A crate can fit up to 8 bottles of water or 20 cans of soft drinks.

Ivan is preparing for a picnic for 40 people and needs to provide at least one bottle of water per person and one can of soft drink for every 2 people.

How many crates does he need?

A 3

B 4

C 5

D 6

E 7

14

12 14

The scale shows part of a measuring device. Estimate the value which the arrow indicates.

A 12.45

B 12.90

C 12.95

D 13.05

E 13.30

15 Express trains from Parramatta to Redfern station take about 27 minutes.

Jenny has a dance class at 5pm.

If it takes her 14 minutes to walk from Redfern station to the studio, which of the following trains would be ideal for her to take from Parramatta?

A 3:59 pm
B 4:13 pm
C 4:20 pm
D 4:25 pm
E 4:45 pm

16 Here is a map of a school.

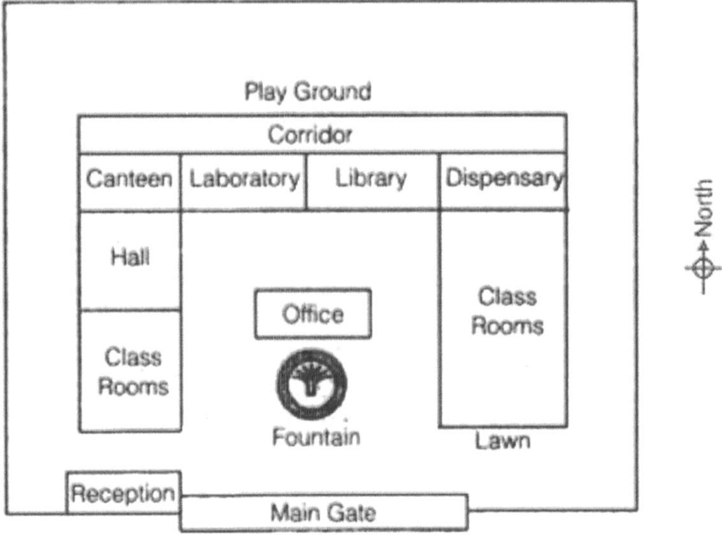

Adrian walks from the canteen to the office. In what direction does Adrian walk?

A north
B south-west
C north-east
D north-west
E south-east

17 Maria was training for a race.

She ran 3 km each day except on Sundays when she ran 5 km.

She ran every day for 31 days, and started her first run on a Friday.

How many kilometres did she run in total?

A 90
B 101
C 103
D 135
E 155

18 Given a three-digit number, a new four-digit number is formed by adding a digit 4 at the end. The new number is then

A the old number plus 4.
B ten times the old number plus 4.
C four hundred plus the old number.
D one hundred times the old number, plus four
E one thousand times the old number, plus four

19 Sasha made a pattern of shapes using black and white blocks.

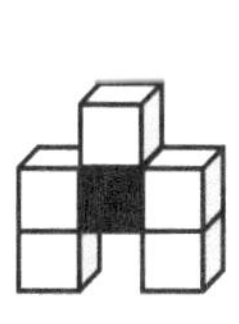

black = 1
white = 5

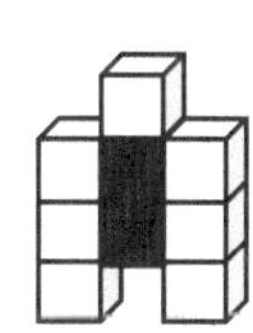

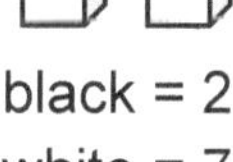

black = 2
white = 7

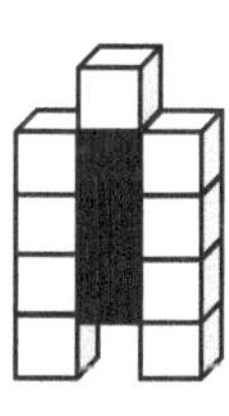

black = 3
white = 9

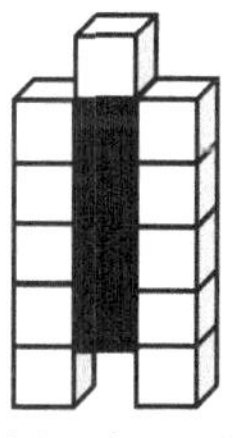

black = 4
white = 11

Sasha made a shape in the pattern that uses 25 white blocks.

How many black blocks did she use?

A 5
B 11
C 15
D 18
E 21

20 Matilda stood in a row of 13 children, where there was a boy between each girl.

If she had 2 girls to the left of her, how many boys were to the right of her?

A 3
B 4
C 6
D 8
E 11

21 What is the least number of squares that must be shaded so that this grid design has an axis of horizontal symmetry?

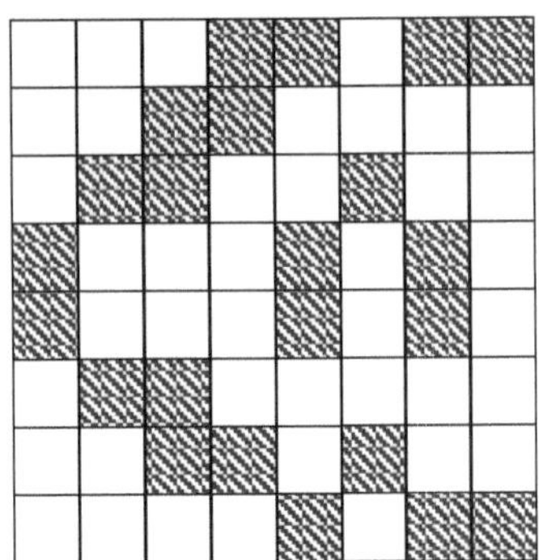

A 1
B 2
C 3
D 4
E 5

22 What will be the next number in this sequence?

4 7 11 18 29 47 ?

A 58
B 60
C 65
D 70
E 76

23 Between Joey, Nicolette and Riya, they have 4 pens, 6 erasers and 5 pencils.

Riya has 4 erasers, 2 more than Nicolette.

Joey has no pens and Nicolette has no pencils.

Joey has the same number of pencils as Nicolette has erasers and Riya has the same number of pens as she has pencils.

How many pens does Nicolette have?

A 1

B 2

C 3

D 4

E 5

24 Kathy was doing her homework when she smudged a question by accident.

Now all she can see is that 8☐6 divided by 7 equals 128. What is the missing number?

A 6

B 9

C 2

D 1

E 0

25 Chloe is saving up for a trench coat from Paris which costs $700.

If she has saved 80% of the required amount, how much more must she save?

A $100.00

B $560.00

C $154.00

D $140.00

E $250.00

26 Which of the following is correct?

A $20 \div 4 \times 6 < 49 \div 7$

B $16 \div 2 \times \frac{27}{3} < 4 \times 16$

C $21 \div 3 \times 5 > 6 \times 5$

D $\frac{18}{2} + 24 \div 3 = 8 \times 2$

E $2 \times 3 \times 5 > 8 \times 4$

27 A pole stands 3.6m high above the ground.

If a quarter of the pole is underground, what is the total length of the pole?

A 3.8 m

B 4.2 m

C 4.8 m

D 5.0 m

E 5.2 m

28 These numbers follow the same rule.

1.5	2.5	2

1.4	2.8	2.1

2.9	?	3

Choose the number that completes the last box.

A 3.1

B 3.2

C 3.9

D 4.0

E 4.1

29 Palindromic numbers are the same when read forwards or backwards.

Here are some palindromic numbers.

11	505	2662

Emma wrote a 5-digit palindromic number.

Its first and last digits add up to 8. The rest of its digits add up to 7.

How many different numbers could Emma have written?

A 1
B 2
C 3
D 4
E 5

30 Find the value of the ○.

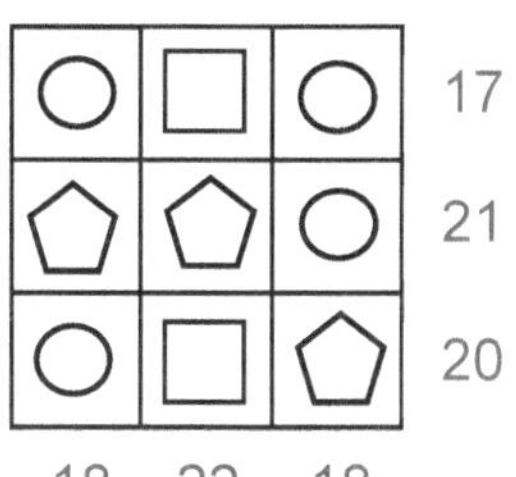

A 5
B 6
C 7
D 8
E 9

31 Dustin is on the university rowing team.

He knows that for every four strokes, he can row the boat 6m.

If he counted 1200 strokes, how far did he row the boat?

A 7200 m
B 1800 km
C 1.5 km
D 1.8 km
E 2.1 km

32 $4 + 2 \times 23 = 2500 \div$ ____

A 5
B 25
C 50
D 10
E 55

33 Four children ate some berries.

Ben and Nina ate half of the berries.
The other half was eaten by Ali and Sarah.

Nina ate twice as many berries as Ben.
Ali ate half as many berries as Ben.

Sarah ate 10 berries.

How many berries were there to start with?

A 40
B 24
C 20
D 12
E 10

34 The average test score for a class of 9 students is 70 out of 100.

A new student joins the class.

If the teacher wanted the class average to be 72, what score must the new student achieve?

A 94
B 92
C 90
D 84
E 82

35 Michael has five cards with numbers on them.

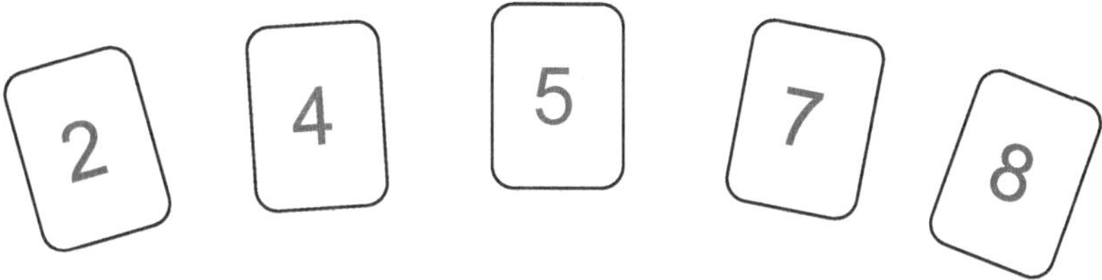

He picks two cards and adds the numbers on them to get a total.

How many different totals can Michael make?

A 6
B 8
C 10
D 12
E 15

OC Practice Test

Mathematical Reasoning 7 (Time allowed: 40 min)

INSTRUCTIONS

1. Write your Name on the cover page.
2. There are 35 questions in this paper. For each question there are five possible answers, A, B, C, D and E. Choose the one correct answer and record your choice on the separate answer sheet. If you make a mistake, erase thoroughly and try again.
3. You will not lose marks for incorrect answers, so you should attempt all 35 questions
4. You must complete the answer sheet within the time limit. There will not be any extra time at the end of the exam to record your answers on the answer sheet.
5. You can use the question paper for working out, but no extra paper is allowed.

Name: ______________________________

1 The graph shows how many students picked Debating, Music bands or Sports as an activity.

Debating	☺ ☺ ☺ ☺ ☺ ☺ ☺
Music bands	☺ ☺ ☺ ☺ ☺
Sports	☺ ☺ ☺ ☺ ☺ ☺ ☺ ☺ ☺

Key:
☺ Represents 3 students

How many students picked Debating?

A 7
B 9
C 18
D 21
E 28

2 Which of these shows the largest angle?

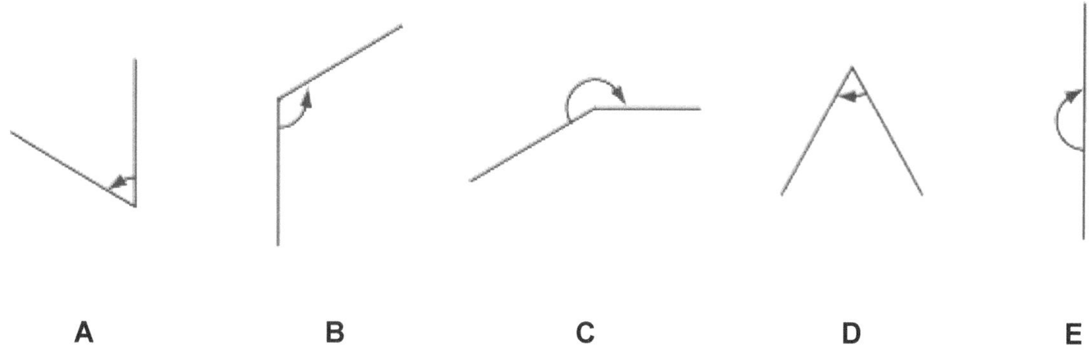

3 What will the next number be in this series of numbers?

1, 2, 6, 15, 31, 56, ?

A 92
B 96
C 100
D 104
E 120

4 If $* + * + ! = 21$,

$? + ? + ? + ! = 18$

and $! = ? + ? + ?$,

then $* =$

A 5
B 6
C 7
D 8
E 9

5 Nick bought 6 packets of tennis balls.

Each packet had 5 balls in it. Nick lost 7 tennis balls.

How can he work out how many tennis balls he has left?

A $6 \times 5 - 7$
B $7 - 6 \times 5$
C $7 \times 5 - 6$
D $6 \times 7 - 5$
E $7 \times 6 + 5$

6 What is the perimeter of this shape?

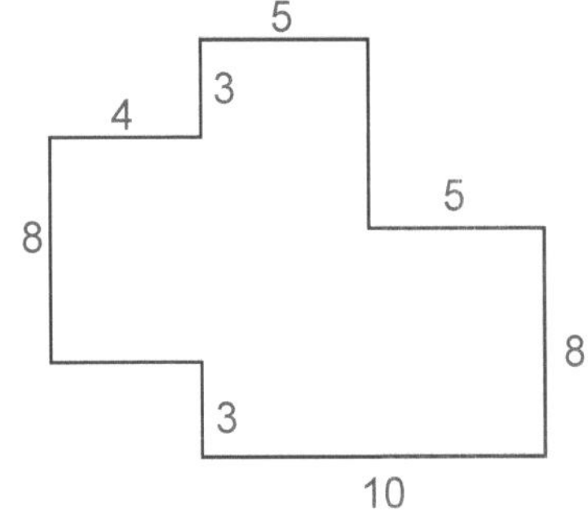

A 54
B 56
C 58
D 60
E 62

7 Anderson had 125 chocolates.

If he eats 10 more chocolates a day than the previous day, and finishes all the chocolates in 5 days, how many did he eat on the third day?

A 25

B 30

C 35

D 40

E 45

8 John has these number cards.

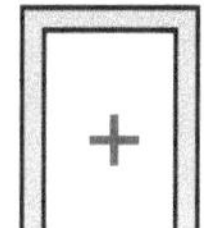

He can arrange them to make a sum. This sum makes a total of 99.

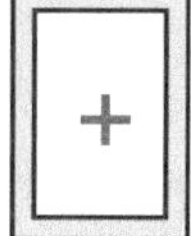

What is the highest sum John can make with his cards?

A 194

B 397

C 693

D 741

E 864

9 A machine packs 4422 toy cars into 11 boxes.

It puts an equal number of cars into each box.

How many toy cars go into one box?

A 42

B 402

C 2211

D 4020

E 48642

10 My alarm clock goes off 360 times in 3 minutes.

How many times does it ring per second?

A 2
B 3
C 4
D 5
E 10

11 What is a hundredth of 45678?

A 45678
B 456.78
C 4.5678
D 45.678
E 4567.8

12 Three-sevenths of the 210 motorbikes in a factory had an engine defect, and only $\frac{1}{3}$ of the remaining bikes were in perfect working condition.

How many in total were in perfect working condition?

A 40
B 30
C 60
D 100
E 80

13 A museum recorded the age and country of its visitors on one day.

		Age group		
		Under 20	20 - 40	Over 40
Country	France	3	7	6
	Canada	2	5	2
	Italy	4	6	5
	Mexico	5	3	7

How many visitors over 20 were from Italy?

A 5

B 6

C 11

D 12

E 13

14 Dennis bought a deck of 60 cards for $24.

He then sold them at six for $3.50.

How much profit did he make?

A $9.00

B $9.50

C $10.00

D $10.50

E $11.00

15 5 people were running in a race.

Sumal is 6 seconds faster than Arun who is 10 seconds faster than Jay.

Amith is 5 seconds slower than Jay.

Nikhil is 8 seconds faster than Jay.

How many seconds slower than Sumal was Nikhil?

A 12
B 10
C 9
D 8
E 6

16 Ashlee sleeps from 10 pm until 6 am every night of the week.

How many minutes does Ashlee sleep in a week?

A 56
B 240
C 480
D 1680
E 3360

17 Jackie can make 2 pies in 3 minutes.

Lenny can make 3 pies in 2 minutes.

How many pies can they make together in an hour?

A 50
B 100
C 130
D 150
E 180

18 $(7 \times 1000) + (3 \times 100) + (4 \times 10) + \frac{2}{10} + \frac{5}{100} =$

A 7340.25
B 7034.25
C 7430.025
D 7430.25
E 7304.25

19 In a bike race Alice finished seven metres ahead of John and three metres behind Cate.

Drew was 4 meters ahead of Alice, but six metres behind of Tina who was three metres behind Sarah.

Who came fourth?

A Alice
B John
C Cate
D Drew
E Sarah

20 Between 11:45 am and 12:00 pm, how many degrees will the minute hand move?

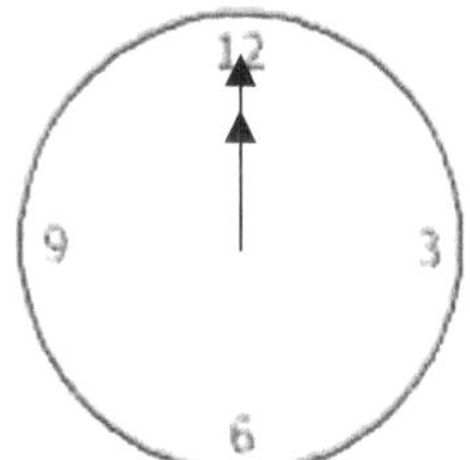

A 60°
B 90°
C 85°
D 120°
E 180°

21 Twenty people were surveyed at a sports centre.

Five people played both squash and tennis.

Eight people did not play squash.

Eleven people did not play tennis.

How many people did not play either squash or tennis?

A 1
B 4
C 10
D 15
E 19

22 If 2 days before tomorrow is Friday, what is 8 days from today?

A Monday
B Tuesday
C Sunday
D Thursday
E Friday

23 Here is a pattern with counters.

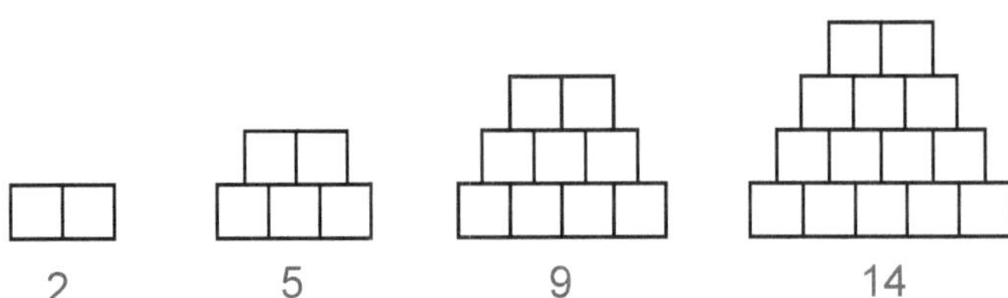

How many counters will be in the next shape of the pattern?

A 19
B 20
C 21
D 22
E 23

24 Of 5 consecutive numbers, the middle 3 numbers add to 15.

What is the 4th consecutive number starting from smallest to largest?

A 10
B 9
C 8
D 7
E 6

25 Q & T below are made of the same sized cubes.

If the value of Q is 5.

What is the value of T?

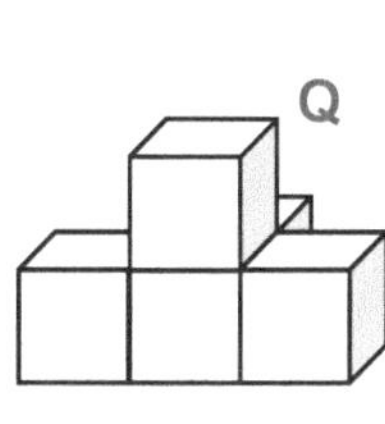

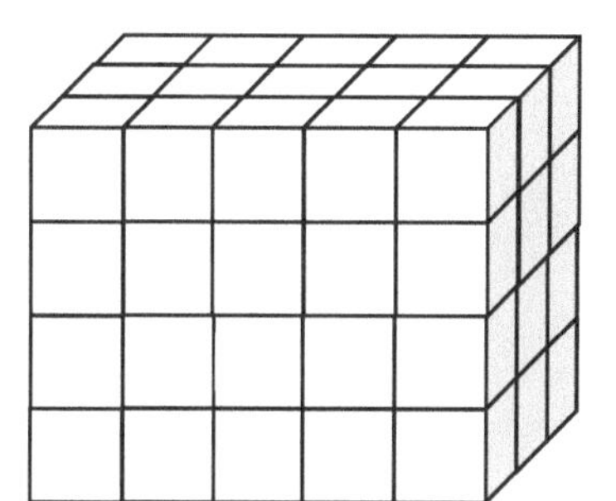

A 60
B 76
C 84
D 105
E 168

26 A + B + C = 30

A is a third of C which is half of B.

What is A + B?

A 12
B 19
C 21
D 24
E 30

27 Charlie made some chocolate chip cookies.

Charlie used 20 chocolate chips for every 100 grams of cookie mixture.

How many chocolate chips did Charlie use for 760 grams of cookie mixture?

A 5
B 38
C 120
D 152
E 160

28 The value of 9 in 7681.29 is

A 9×0.01
B 9×10
C 9×0.1
D 9×0.001
E 9×1

29 What is the third largest number you can make using the first four even numbers and zero?

A 86042
B 86420
C 86402
D 86240
E 86204

30 What fraction of this figure is shaded?

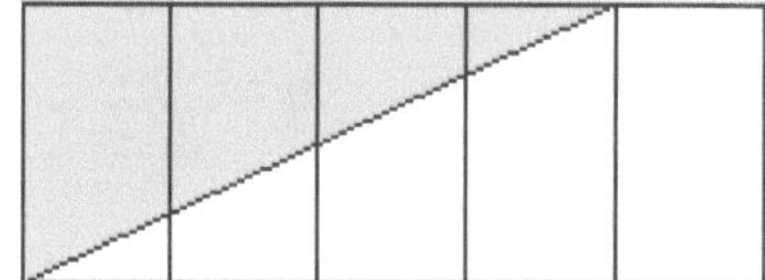

A $\frac{2}{5}$

B $\frac{1}{2}$

C $\frac{6}{10}$

D $\frac{1}{4}$

E $\frac{3}{5}$

31 2 burgers + 1 drink cost $10.50

1 burger + 2 drinks cost $7.50

How much does it cost for one burger and one drink?

A $10.00

B $7.00

C $9.00

D $8.00

E $6.00

32 Which number replaces the question mark?

2 3 5 7 ? 13 17 19

A 10
B 9
C 11
D 8
E 12

33 The picture shows the area around Tony's swimming pool.

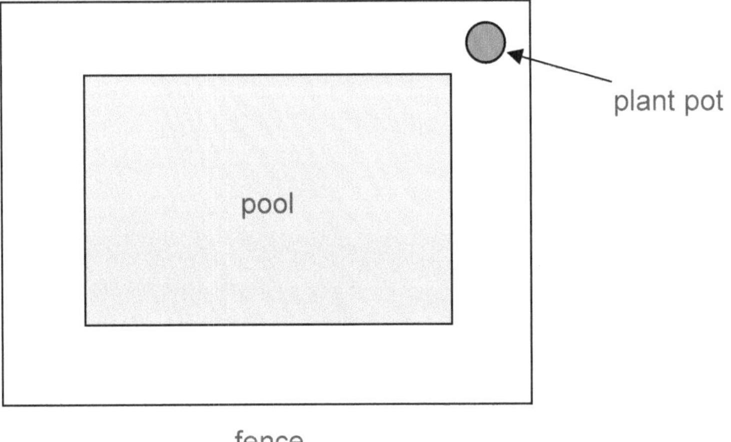

He plans to have 7 pots along each side of the pool including a pot at every corner. How many pots would Tony need?

A 20
B 24
C 28
D 30
E 32

34 Harry has some jars of honey. The jars come in three sizes.

Harry has at least one of each size of jar and all his jars are full.

Altogether he has 3 kilograms of honey in the jars.

How many jars does he have?

A 5
B 6
C 7
D 8
E 9

35 The bike Alan is saving for costs $600.00.

If Alan has saved 60% of the needed amount, how much more must he save?

A $360.00
B $120.00
C $180.00
D $60.00
E $240.00

OC Practice Test

Mathematical Reasoning 8 (Time allowed: 40 min)

INSTRUCTIONS

1 Write your Name on the cover page.

2 There are 35 questions in this paper. For each question there are five possible answers, A, B, C, D and E. Choose the one correct answer and record your choice on the separate answer sheet. If you make a mistake, erase thoroughly and try again.

3 You will not lose marks for incorrect answers, so you should attempt all 35 questions

4 You must complete the answer sheet within the time limit. There will not be any extra time at the end of the exam to record your answers on the answer sheet.

5 You can use the question paper for working out, but no extra paper is allowed.

Name: ______________________________

1 Olivia is in box S1.

Who is in box U2

A Emma
B Lucas
C Liam
D Henry
E Mia

2 How many of these shapes have at least one right angle?

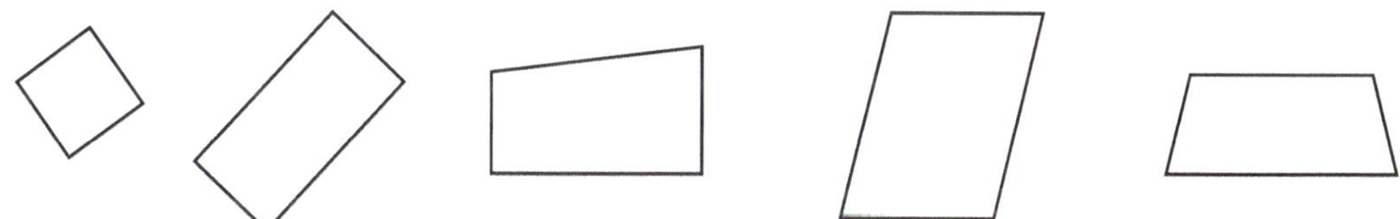

A 1
B 2
C 3
D 4
E 5

3 Justin's mass is 30 kg less than Mac's but 15 kg more than Jim's.

If Mac's mass is twice his brother's mass of 62 kg, how much heavier is Mac than Jim?

A 30 kg
B 45 kg
C 79 kg
D 96 kg
E 124 kg

4 What is the smallest number that can be divided by 3, 4 and 5 with no remainder?

A 15
B 45
C 60
D 80
E 120

5 Jemma can walk 80 metres in 1 minute. How far can she walk in 5 minutes?

A 16 metres
B 85 metres
C 160 metres
D 320 metres
E 400 metres

6 Twenty-one hundred bikes were expected to be sold at a store over a month, however almost double were sold by the end of the month.

Which answer shows how many bikes were sold?

A 4202
B 4199
C 2000
D 4230
E 3890

7 This 3D shape has a value of 5.

What is the value of this 3D figure?

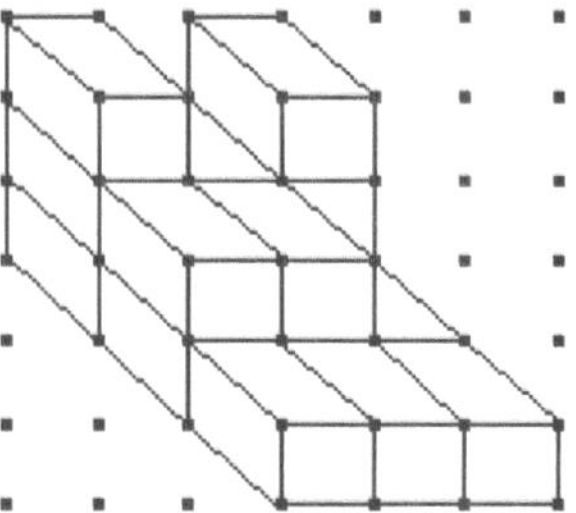

A 40
B 45
C 50
D 55
E 60

8 Julia is facing north. She turns 90° to her right, then turns 180° to her left.

What direction is she now facing?

A northeast
B east
C south
D west
E northwest

Use the diagram below to answer questions 9 and 10.

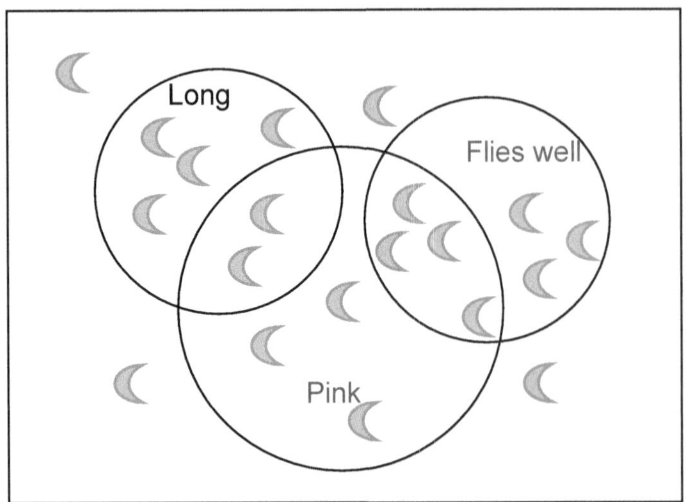

Each ☾ represents 3 boomerangs, which can either be pink or brown, flies well or do not fly well, and are long or short.

9 How many boomerangs are pink, but do not fly well?

A 5
B 10
C 13
D 15
E 18

10 How many boomerangs are not pink, and short?

A 12
B 27
C 24
D 21
E 18

11 Which two pictures are three-quarter shaded?

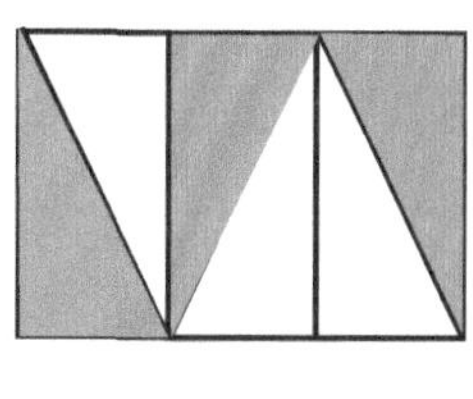

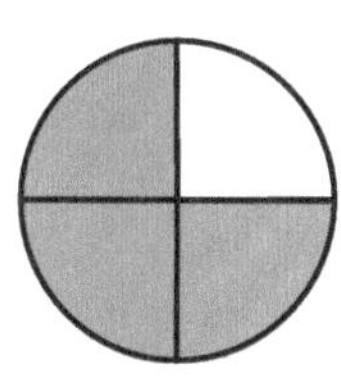

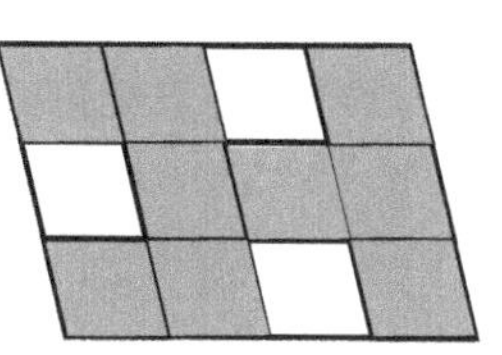

W **X** **Y** **Z**

A W and Y

B W and Z

C W and X

D X and Y

E X and Z

12 In a theatre, Indra is sitting in the 9th row from the front.

This row is also the 15th row from the back.

How many rows of seats are there in this theatre?

A 6

B 23

C 24

D 25

E 26

13 Ivan has 6 pieces of string. Each piece of string is 25 cm.

He ties the strings together to make one string that is 90 cm long.

He uses the same amount of string for each knot.

How much string does he use in one knot?

A 10 cm
B 12 cm
C 15 cm
D 18 cm
E 20 cm

14 Two apples and an orange cost $3.00.

An apple and two oranges cost $3.30.

How much is an apple?

A $0.90
B $1.00
C $1.10
D $1.20
E $1.30

15 How many minutes are there in five and a half hours?

A 55
B 300
C 330
D 480
E 550

16 Cody has 4 hats, 6 shirts and 2 pairs of pants.

How many combinations of different outfits does he have?

A 48
B 58
C 64
D 72
E 84

17 $1\frac{1}{2}$ bottles of soda are required to fill one jug.

For his birthday party, Jim needs 12 jugs filled.

How many bottles of soda must he buy for his party?

A 18
B 17
C 16
D 15
E 14

18 If the following shapes balance each other like so,

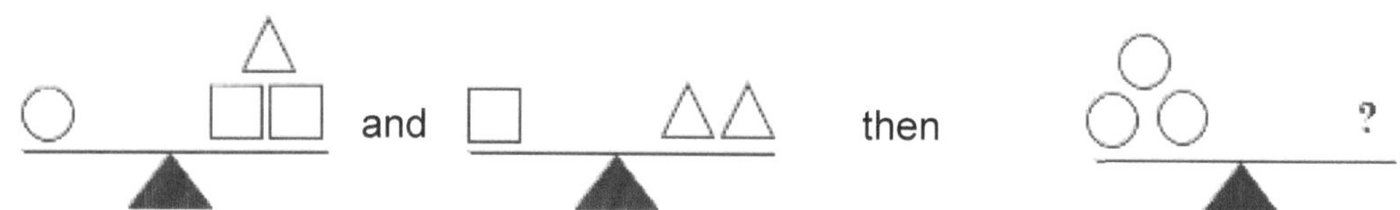

How many triangles are needed to complete the balance?

A 6
B 9
C 12
D 15
E 18

19 The third number of a sequence is 13 and the ninth number is 37.

What is the twelfth number in this sequence?

A 45
B 49
C 51
D 53
E 55

20 The table shows the different ways students travel to school.

Ferry	Bus	Train	Car	Walk
6	12	4	8	10

Which section of the pie graph represents the number of students who catch the train?

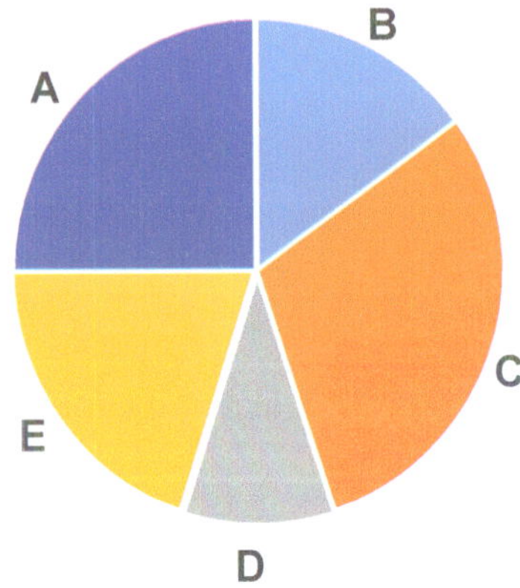

21 Chris rode his skateboard down a ramp with 3 segments.

It took him 20 seconds to reach the end of the first segment, each following segment took half the time of the last.

How long did it take the skateboard to reach the bottom?

A 50 seconds
B 45 seconds
C 40 seconds
D 35 seconds
E 30 seconds

22 For a concrete mix to fill a driveway, Bob needs 4 parts aggregate, 2 parts sand and 1 part of both cement and water.

His driveway measures twelve metres by fifteen metres, and is 20 centimetres deep.

How much sand does he require?

A 5.14 m^3
B 1028 cm^3
C 900 cm^3
D 90000 cm^3
E 9 m^3

23 Which number will come next in this number pattern?

1 2 4 8 16 32 64

A 128
B 136
C 184
D 156
E 256

24 Which of these statements is/are correct?

X $\frac{2}{5} + \frac{2}{5}$ is less than $\frac{3}{5}$

Y $1 - \frac{1}{3}$ is more than $\frac{1}{3}$

Z $\frac{2}{4} \times \frac{1}{2}$ is more than $\frac{1}{8}$

A statement X only
B statement Y only
C statement Z only
D statements X and Y only
E statements Y and Z only

25 A baby crawled 4 metres in 10 seconds.

At what speed did the baby crawl?

A 4 centimetres per second
B 25 centimetres per second
C 40 centimetres per second
D 60 centimetres per second
E 100 centimetres per second

26 What is the difference between 33.3 and 3.33?

A 29.97
B 30.97
C 30.00
D 30.03
E 36.63

27 What is the area of the following two shapes?

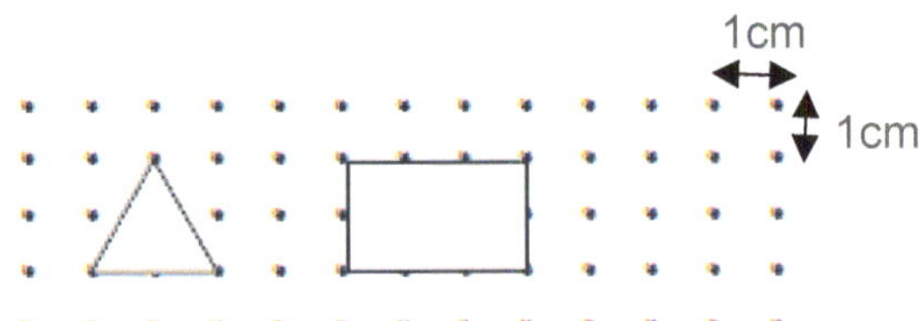

A 14 cm^2
B 12 cm^2
C 10 cm^2
D 8 cm^2
E 6 cm^2

28 Find the average of the digits in the quotient of 3948 and 7.

A 2
B 5
C 7
D 30
E 564

29 In a factory there were originally 50 more in number of cars than motorbikes.

During the year, 100 cars were sold, whilst 220 motorbikes were manufactured.

There are now 420 motorbikes.

How many cars are now in the factory?

A 300
B 250
C 350
D 150
E 100

30 The road connecting to towns A & B is 12.4 km long and is divided into four sections.

If the first three sections are 1.3 km, 3.2 km and 3.6 km long respectively, what is the length of the last section of road?

A 3.9 km
B 4.1 km
C 4.3 km
D 4.6 km
E 4.9 km

31 This shape was made from 5 squares each with an area of 16 cm^2.

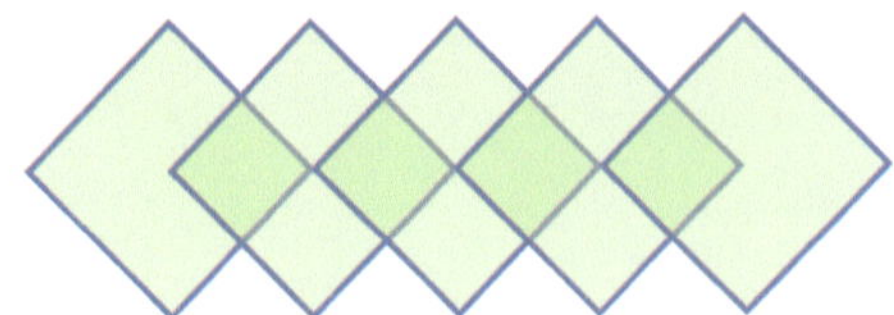

What is the area of the shape?

A 60 cm^2

B 64 cm^2

C 68 cm^2

D 72 cm^2

E 80 cm^2

32 The picture shows five blocks made out of cubes. Each block is a rectangular prism.

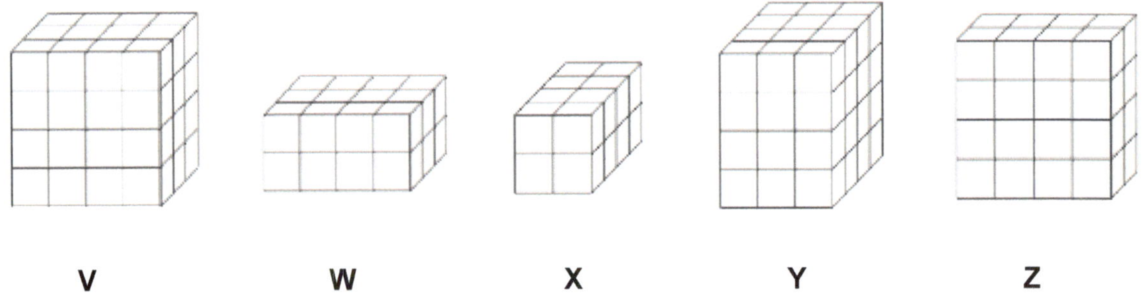

Which two blocks have the same volume?

A X and Z

B W and Y

C X and Y

D W and Z

E V and Y

33 Harry has a spinner that is split into five equal sections.

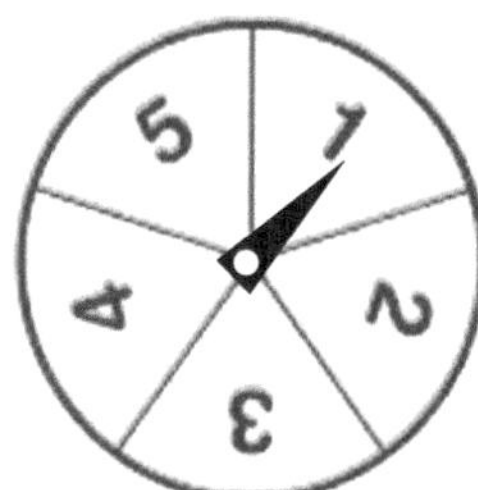

He spins the spinner and it lands on 1.

Now Jayden is going to spin the spinner once. Which of these statements are correct?

X The probability of Jayden getting a 1 is 20%
Y The probability that Jayden's number is more than Harry's is 80%
Z The probability that Jayden and Harry's scores add up to make more than 6 is 50%

A none of them
B statements X and Y only
C statements X and Z only
D statements Y and Z only
E statements X, Y and Z

34 Mr Harrison asked the students in his class to choose either cricket or soccer for sport. This table shows the results.

	Cricket	Soccer
Girls	6	13
Boys	8	9

How many students are there in this class?

A 36
B 29
C 22
D 17
E 14

35 A cook uses $\frac{3}{5}$ kg of meat to make 4 hamburgers.

How many kg of meat does he need to make 80 hamburgers?

A 28

B 24

C 20

D 15

E 12

OC Practice Test

Mathematical Reasoning 9 (Time allowed: 40 min)

INSTRUCTIONS

1 Write your Name on the cover page.

2 There are 35 questions in this paper. For each question there are five possible answers, A, B, C, D and E. Choose the one correct answer and record your choice on the separate answer sheet. If you make a mistake, erase thoroughly and try again.

3 You will not lose marks for incorrect answers, so you should attempt all 35 questions

4 You must complete the answer sheet within the time limit. There will not be any extra time at the end of the exam to record your answers on the answer sheet.

5 You can use the question paper for working out, but no extra paper is allowed.

Name: ____

1 The robot cleaner has six hands. Each hand has two fingers.

How many fingers would 15 robot cleaners have?

A 27
B 60
C 95
D 180
E 240

2 $4 + 8 \times 3 = ? - 9$

A 16
B 24
C 31
D 37
E 45

3 The perimeter of the below figure is 80 cm and is made up of identical rectangles.

The widths of each of the rectangles are half of their lengths.

What is the length of each rectangle?

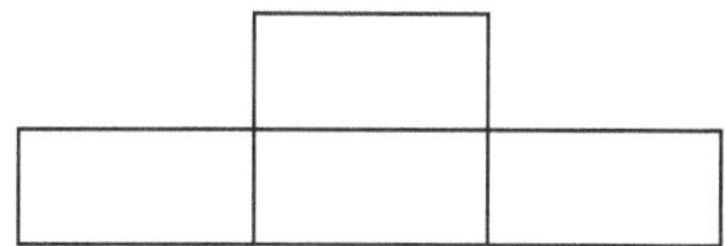

A 4 cm
B 5 cm
C 6 cm
D 8 cm
E 10 cm

4 $\frac{6}{8}$ of a certain number is the same as $\frac{3}{4}$ of 32.

What is the number?

A 32
B 38
C 42
D 48
E 56

5 Dennis, Vivian, Neil, and Andrew were running in a race.

Vivian finished in front of Andrew who finished after Neil.

Which of the following shows the correct order of position if the average of Dennis and Andrew's time is equal to Neil's, but less than Vivian's?

A Vivian, Dennis, Andrew, Neil
B Dennis, Vivian, Neil, Andrew
C Dennis, Vivian, Andrew, Neil
D Dennis, Neil, Vivian, Andrew
E Vivian, Andrew, Dennis, Neil

6 Brian is posting three of these five toys in a parcel.

doll = 85 g   bear = 99 g   car = 46 g   ball = 65 g   train = 77 g

The parcel has to be **less** than 200 grams.

Which three toys together weigh closest to 200 grams?

A doll, bear, ball
B doll, car, ball
C doll, bear, train
D bear, car, ball
E bear, car, train

7 These number groups follow the same pattern.

Which number best replaces the question mark?

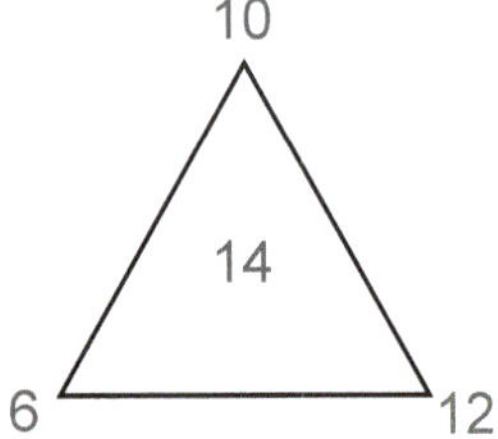

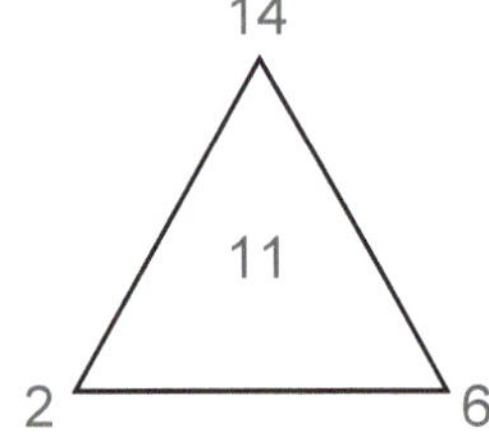

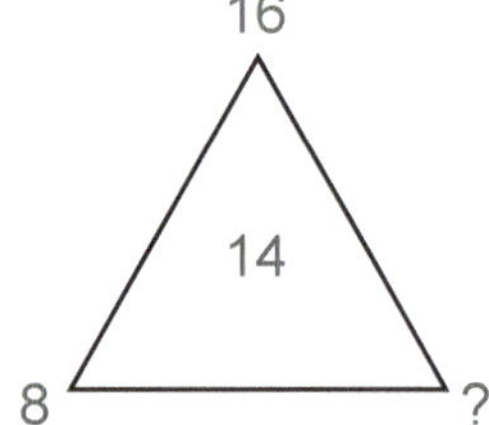

A 4
B 8
C 12
D 16
E 20

8 I have a poster that is 2m long and 1m wide.

I want to place it in the centre of my ceiling which is 5m long by 4m wide.

How far away in total are the edges on the poster from the edges of the ceiling?

A 5 m
B 6 m
C 7 m
D 8 m
E 9 m

9 3 Friends are seated around a circular table.

They are numbered 1- 3.

How many ways can they be seated?

A 1
B 2
C 3
D 4
E 5

10 Here are some shapes.

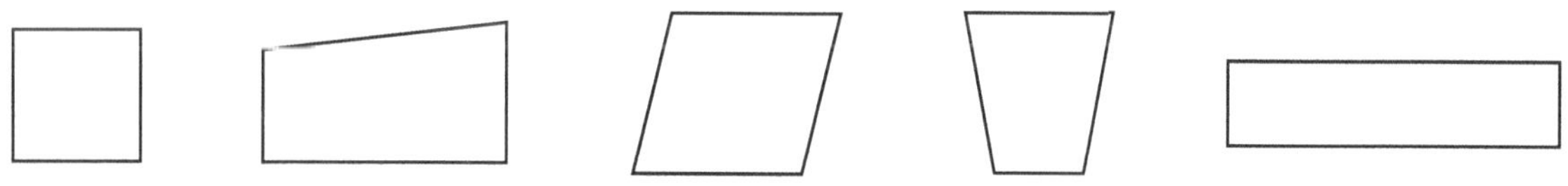

How many of these shapes are parallelograms?

A 1
B 2
C 3
D 4
E 5

11 Last year Jessica was a third of her dad's age.

Her mum is now 40 and 3 years older than her husband.

How old will Jessica be in three years' time?

- **A** 16 years
- **B** 17 years
- **C** 18 years
- **D** 19 years
- **E** 20 years

12 Using these triangles, (2cm, 2cm) how many can fit inside this rectangle?

- **A** 10
- **B** 12
- **C** 16
- **D** 20
- **E** 28

13 Students may play football, cricket or both sports at lunch time.

The table shows the activities of 30 students at lunch time last Thursday.

Activities	
football	15
cricket	15
neither sport	5

How many students played both football and cricket?

A 1
B 2
C 3
D 4
E 5

14 I want to make a rectangular prism by adding an extra layer of small blocks to all sides of this 3D figure.

How many additional blocks do I need?

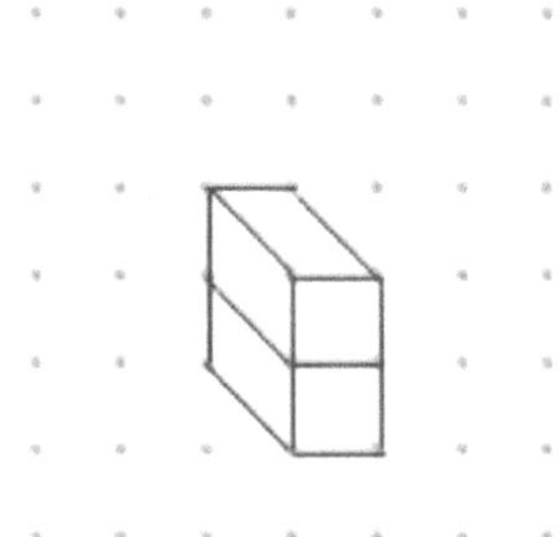

A 32
B 34
C 44
D 52
E 60

15 Six identical boxes weigh 42 kg.

How much would 10 boxes weigh?

A 60 kg
B 70 kg
C 80 kg
D 85 kg
E 90 kg

16 How many triangles are there in this shape?

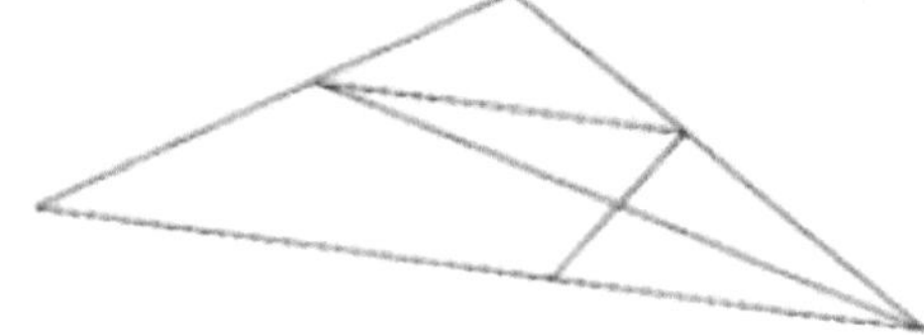

A 4
B 5
C 6
D 7
E 9

17 Krishni stole 25 cookies from the cookie jar.

For every 4 she took, she gave one to her friend Emily.

How many cookies did Emily get?

A 2
B 3
C 4
D 5
E 6

18 Jonathan sat three thinking skills tests and got 80% of all questions correct.

Here are his results:

Test 1	Test 2	Test 3
$\frac{28}{40}$	$\frac{30}{40}$	$\frac{\square}{40}$

How many questions did Jonathan get correct in Test 3?

A 40
B 38
C 36
D 34
E 32

19 When Jimothy parked his car at university, there were 10 to the right of his and two to the left.

If his car is now in the middle of 25 cars, how many **more** must have parked to his left?

A 8
B 9
C 10
D 12
E 13

20 Every time Joel washes his car, the tap to which the hose is connected leaks 5L.

Joel has a bucket that holds 2.5L.

If Joel washes his car 4 times a week, how many buckets of water does he end up collecting?

A 8 buckets
B 10 buckets
C 12.5 buckets
D 15 buckets
E 20 buckets

21 When a class is divided into groups of three, there is one person left over, However when it is divided into groups of 8, there are 4 people left over.

How many are in the class?

A 20
B 24
C 28
D 32
E 36

22 Manling was facing south-west and then she turned 90 degrees clockwise.

Which direction was she then facing?

A south-east
B north-east
C south
D north
E north-west

23 Jeremy cut a long piece of rope six times and then divided each piece into 3 parts.

How many pieces of rope did he finish with?

A 21
B 18
C 24
D 28
E 15

24 Choose the number that is missing from this sequence.

2 3 4 6 6 9 8 12 10 ? 12

A 11
B 13
C 15
D 17
E 19

25 What 3 numbers is 30 the average of?

A 31, 32, 33
B 29, 28, 32
C 29, 30, 31
D 30, 31, 32
E 28, 31, 34

26

then ◇ = ?

A 9 kg
B 11 kg
C 13 kg
D 14 kg
E 15 kg

27 One step taken by George is 4 of Arthur's steps.

One step taken by Nick is three times as long as Arthur's.

If Arthur's steps are 20 centimetres long each and George takes 15 steps, how many metres will Nick have walked if he takes the same amount of steps as George?

A 9

B 3

C 1

D 5

E 12

28 Use the graph to answer the question. It shows the height and age of four people.

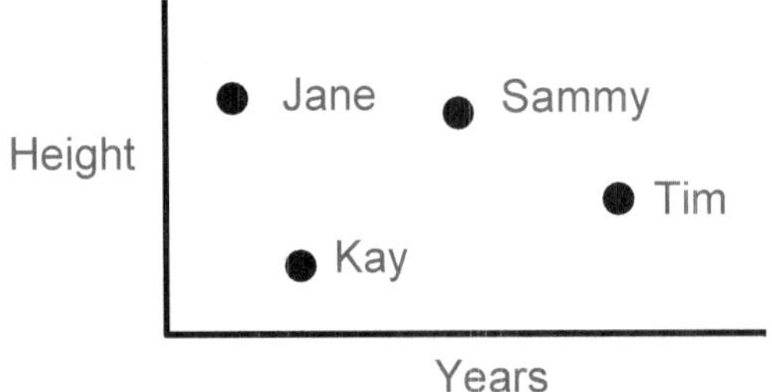

Choose the most correct statement.

A Kay is really short for her age.

B Tim is average for his height.

C Sammy is really tall for her age.

D Jane is really tall for her age.

E Tim is taller than Jane.

29 A piece of timber was 3.2m long.

Four equal lengths of 50cm were cut from it.

What length of timber remained?

A 2.2 m
B 1.5 m
C 1.2 m
D 1.55 m
E 10.5 m

30 There are 50 children in Year 6.

Of these, 60% will be going to Baulkham Hills High School next year.

How many will be going elsewhere?

A 30
B 28
C 25
D 22
E 20

Use the diagram below to answer questions 31 and 32.

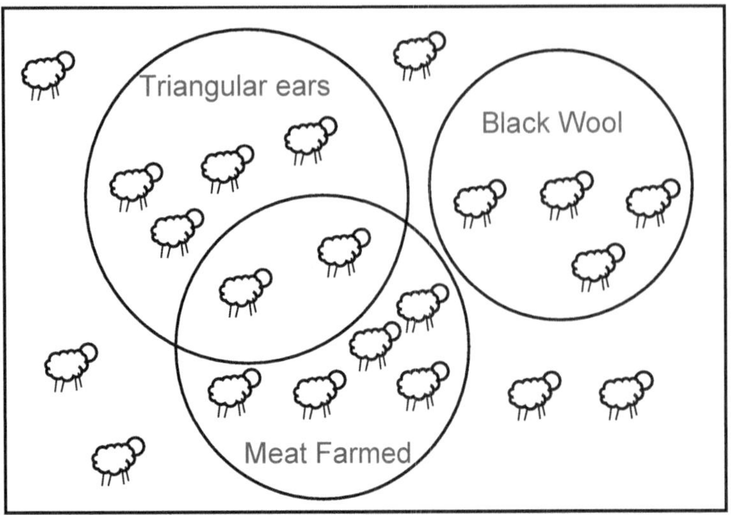

Each represents five sheep.

Sheep either have black or white wool.

They are either farmed for their meat or their wool, and have either triangular or rounded ears.

31 How many sheep are on this farm?

A 21
B 22
C 105
D 110
E 115

32 How many sheep with rounded ears are farmed for their wool?

A 10
B 20
C 50
D 55
E 100

33 If Andrew is playing Maple-story and has $\frac{3}{5}$ of total points until levelling up.

If he has scored 20 experience points so far, how many points in total would it take to clear the level from start to finish?

A 20 experience points
B 30 experience points
C 40 experience points
D 50 experience points
E 60 experience points

34 For her children at school, Mrs Jane bought five textbooks at $8.40 each, one book at $12.40 and two other novels, each costing the same amount.

She spent a total of $75.

How much was one of the two novels?

A $8.50
B $9.80
C $10.30
D $11.60
E $12.40

35 A group of students were asked how many siblings they had at home.

The graph shows the results:

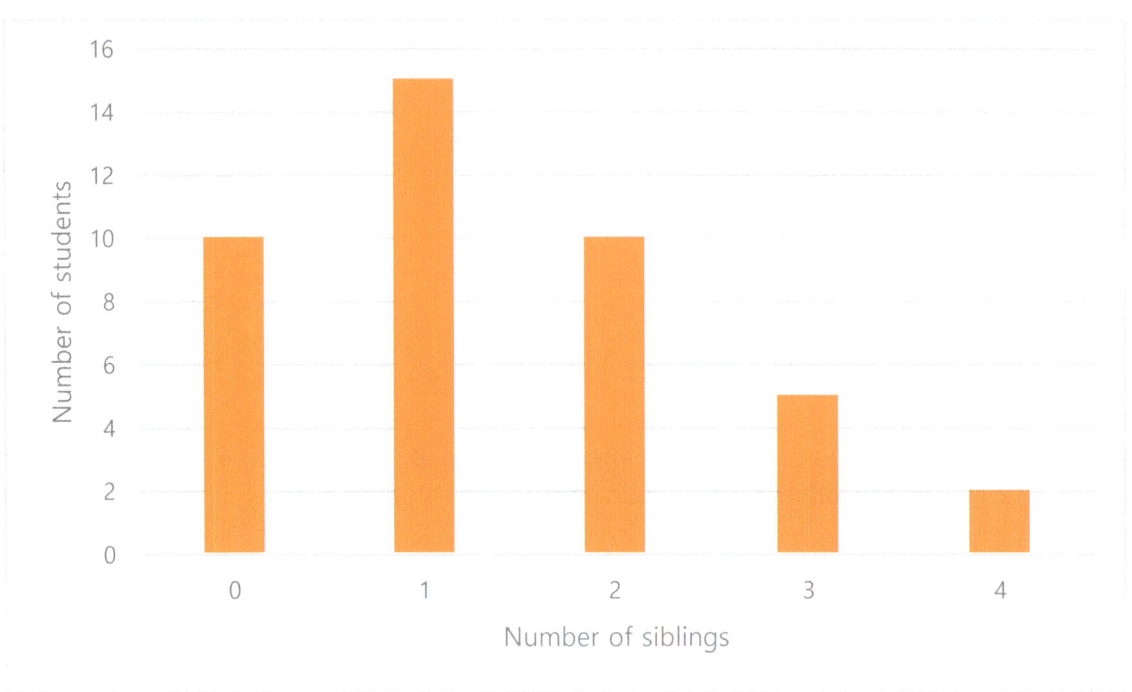

Joanne reads the graph and makes 3 claims:

1 More than 20 students had 3 or 4 siblings.
2 Exactly two times as many children had 2 siblings as 3 siblings.
3 There were 13 more students with 1 sibling than students with 4 siblings.

Using the information on the graph, which of Joanne's claims is/are correct?

A Claim 1 only

B Claim 2 only

C Claim 3 only

D Claims 1 and 2

E Claims 2 and 3

OC Practice Test

Mathematical Reasoning 10 (Time allowed: 40 min)

INSTRUCTIONS

1 Write your Name on the cover page.

2 There are 35 questions in this paper. For each question there are five possible answers, A, B, C, D and E. Choose the one correct answer and record your choice on the separate answer sheet. If you make a mistake, erase thoroughly and try again.

3 You will not lose marks for incorrect answers, so you should attempt all 35 questions

4 You must complete the answer sheet within the time limit. There will not be any extra time at the end of the exam to record your answers on the answer sheet.

5 You can use the question paper for working out, but no extra paper is allowed.

Name: ______________________________

1 Natalia is making cupcakes. She needs 60g of butter to make 12 cupcakes.

What is the largest number of cupcakes that Natalia can make with 120g of butter?

A 10
B 24
C 60
D 120
E 132

2 Which intersecting lines form right angles?

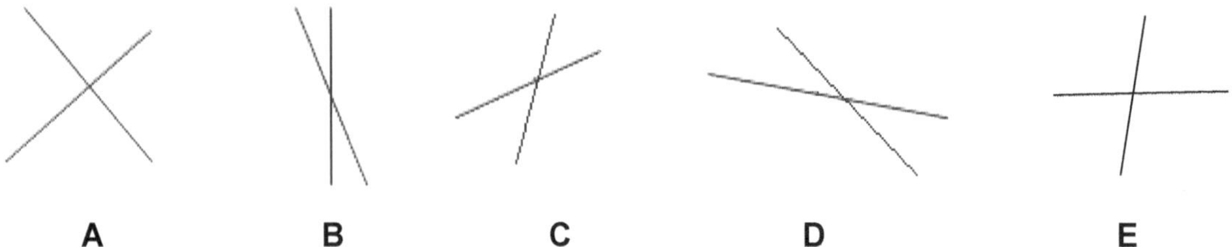

3 Tony has four parrots and five canaries in a cage.

When Tony goes to feed his birds, he notices that one of them is missing.

What is the chance that the missing bird is a canary?

A 1 out of 2
B 1 out of 5
C 4 out of 9
D 5 out of 9
E 9 out of 9

4 Vivian drew five shapes.

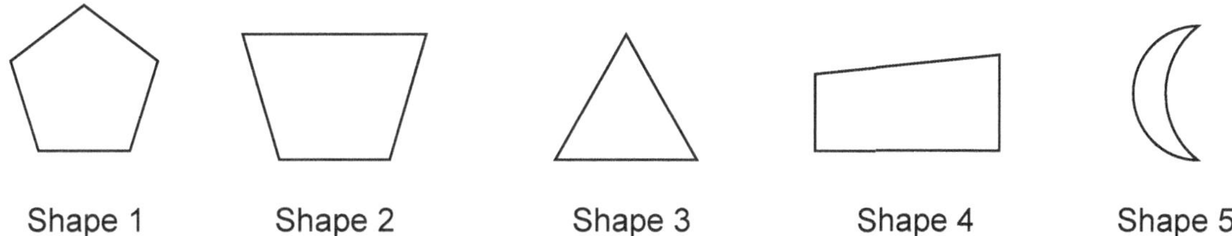

Which shapes have only **one** line of symmetry?

A Shape 3 and Shape 4
B Shape 2 and Shape 3
C Shape 4 and Shape 5
D Shape 1 and Shape 2
E Shape 2 and Shape 5

5 The following graph shows how far 4 athletes could run with how much water they drank.

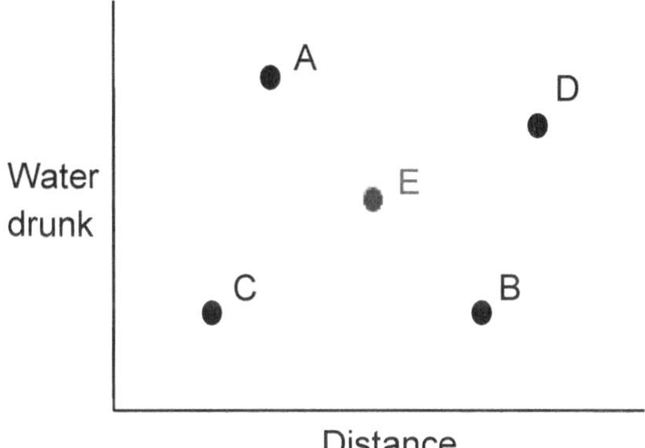

Who was the most efficient athlete?

A A
B B
C C
D D
E E

6 Ms Scott has 38 microscopes.

She needs to have two sets of 28 for her science classes.

How many more microscopes does Ms Scott need?

A 10
B 18
C 20
D 22
E 24

7 This 3D letter P is made out of wood. How many faces does it have?

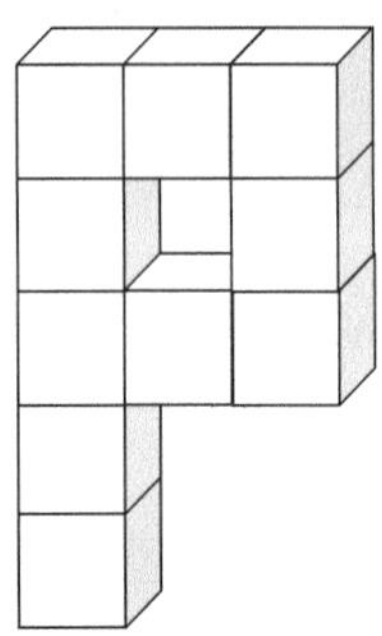

A 8
B 10
C 12
D 14
E 16

8 Calculate the value □

$4 \times 7 - \frac{2}{3} \times 9 = \square$

A 18
B 22
C 30
D $\frac{2}{3}$
E $2\frac{2}{3}$

9

Vivian, Jessica and Mary shared a pizza.

Vivian ate $\frac{1}{8}$ of the pizza, Jessica ate $\frac{3}{8}$ of the pizza and Mary ate the rest.

Which of the following statements is/are correct?

1 Mary ate less than half of the pizza.
2 Jessica ate more than one quarter of the pizza.
3 Vivian and Jessica ate less than half of the pizza altogether.

A statement 1 only

B statement 2 only

C statement 3 only

D statements 1 and 2 only

E statements 1 and 3 only

10 Joshua and Harry work in a supermarket at the checkouts.

When Joshua has served 6 customers, Harry served eight. At this rate, how many customers would Joshua have served when Harry has served his 40th customer?

A 38

B 34

C 36

D 30

E 26

11 Sunil is making a model plane.

He has two wooden sticks of lengths 1.12 metres and 0.88 metres.

He glued the ends of the two sticks together to make a piece for his plane.

What is the length of this piece in metres?

A 2.10
B 2.00
C 1.90
D 1.12
E 2.12

12 How many triangles are there in the shape below?

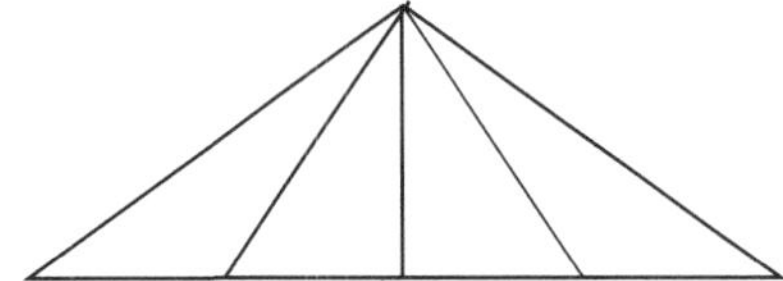

A 4
B 5
C 7
D 8
E 10

13 A local shop is offering one bouncy ball free with every five bouncy balls a customer pays for.

Jess pays the normal price for 30 bouncy balls.

How many bouncy balls will she receive in total?

A 30
B 31
C 35
D 36
E 40

14 A dark square has been placed on four white squares as shown.

The sides of these squares are 4 cm long.

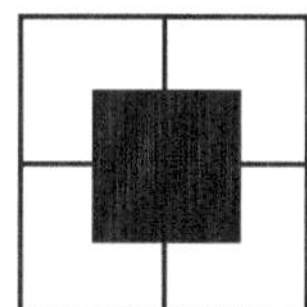

What is the area of the white region surrounding the dark square?

A 16 cm^2

B 24 cm^2

C 48 cm^2

D 56 cm^2

E 64 cm^2

15 5 children equally shared a box of chocolates and 3 children equally shared a box of marshmallows.

Both boxes contained the same number of chocolates and marshmallows.

How many chocolates were there if each child had 6 more marshmallows than a child who had chocolates?

A 60

B 45

C 35

D 75

E 50

16 Henry had a cake for his party.

At the party, $\frac{5}{8}$ of the cake was eaten and leftovers weighed 600 g.

What was the weight of the whole cake?

A 600 g

B 800 g

C 1.2 kg

D 1.6 kg

E 2 kg

17 At lunchtime, a group of friends were eating a large number of grapes.

Knowing that Dennis had 15 more than Jonathan who had 13 less than Vivian, and Neil had 4 less than Dennis who also happened to have 1 more than Andrew.

Who had the third most number of grapes?

A Dennis

B Neil

C Vivian

D Andrew

E Jonathan

18 When folded on the given diagonal, how many pairs of shaded squares will coincide?

A 3
B 4
C 5
D 6
E 7

19 To complete this 3D drawing of a triangular prism, one more pair of coordinates is needed.

Which pair is it?

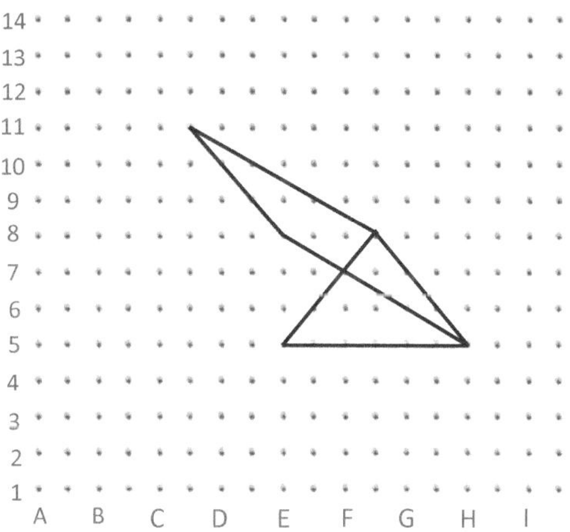

A B8
B C11
C A5
D D12
E E4

20 Mandy was lost. She ran 5 km west, turned right and walked 2 km, turned left and ran 3 km, and turned right and ran 4 km.

How far north from her starting point was she now?

A 8 km

B 6 km

C 5 km

D 3 km

E 2 km

21 Anish was trying to lose weight.

At the beginning of summer, he weighed 76.5 kg.

At the end of summer, he weighed 69.8 kg.

How many kilograms did Anish lose?

A 16.7

B 13.3

C 10.7

D 7.3

E 6.7

22 What number will come next in this sequence?

2, 5, 4, 6, 6, 7, 8, 8, 10, 9, ☐

A 12

B 13

C 10

D 11

E 14

23 Which two 3D shapes will fit together to form a rectangular prism?

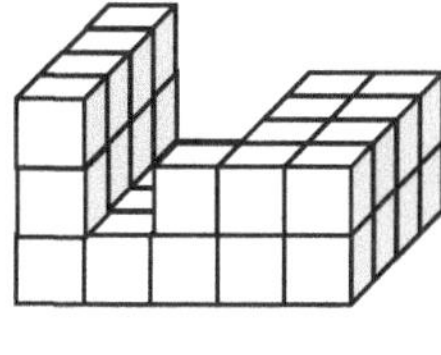
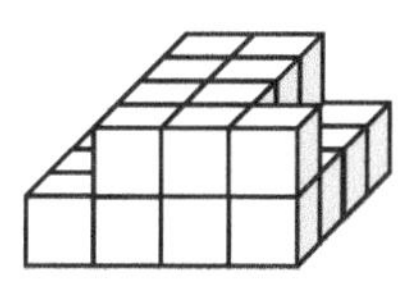
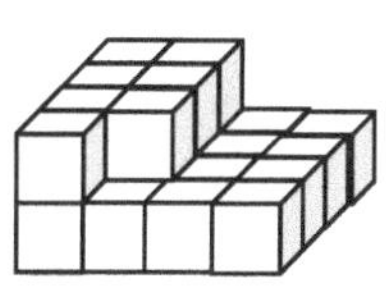
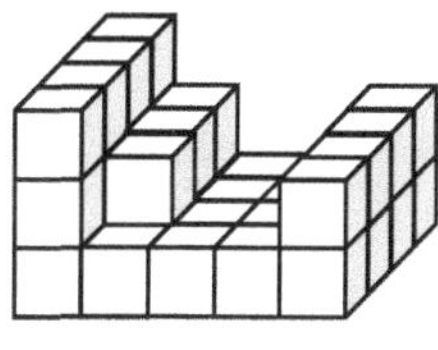

A **B** **C** **D**

A A and C
B A and B
C B and D
D C and D
E C and B

24 Find the value of □ and △.

$$\begin{array}{r} 3\ \square\ \triangle \\ 6\ \times \\ \hline 1\ 9\ 9\ 2 \\ \hline \end{array}$$

A 9, 7
B 7, 2
C 5, 7
D 3, 2
E 4, 7

25 Kathy and her two sisters shared $80.

Her eldest sister received 3 times Kathy's amount and her younger sister received one-quarter of the total.

How much did Kathy get?

A $15.00
B $20.00
C $25.00
D $30.00
E $35.00

26 Sunil made this number pattern:

4, 9, 19, 39 ...

He used this strategy:

4 + 5 = 9
9 + 10 = 19
19 + 20 = 39

Adam made the same pattern using a different strategy.

He started with the number 4.

What is the strategy that Adam used?

A add 10
B multiply by 3 then subtract 3
C double the number then add 1
D write the next number that ends with 9
E add 1 then multiply by 2

27 What is the value of □ ?

$$3 \times \frac{7}{8} = 7 \times \square$$

A $\frac{3}{8}$

B $\frac{4}{9}$

C $\frac{4}{10}$

D $\frac{2}{7}$

E $\frac{2}{3}$

28 Which two squares have the same fraction shaded?

A

B

C

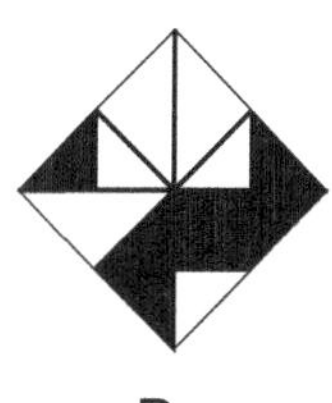
D

A B and C
B C and A
C A and D
D B and D
E C and D

29 Jackie can make 2 pies in 3 minutes.

Lenny can make 2 pies in 2 minutes.

How many pies can they make together in an hour?

A 50
B 90
C 100
D 130
E 150

30 Christie bought a folder for $4.00, six notebooks for $2.00 each, a set of pens for $8.00 and a stapler for $6.00.

She got 10% off her purchase, how much change did she get from $50?

A $18.20
B $20.20
C $23.00
D $26.00
E $28.00

31 Joyce bought a sack of pineapples.

It held 30 pineapples with an average weight of 0.5kg each.

She kept one-third and sold half of the remainder to her neighbour, Michael.

How many kilograms of pineapple did Michael buy?

A 8 kg
B 7 kg
C 6 kg
D 5 kg
E 4 kg

32 Using the number of small blocks used in each figure, which statement is true?

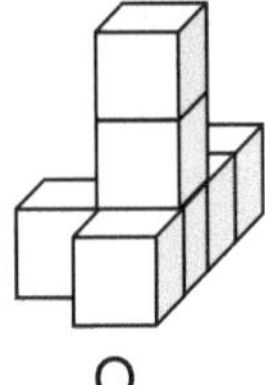
O

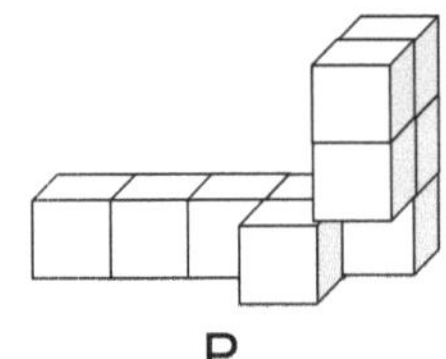
P

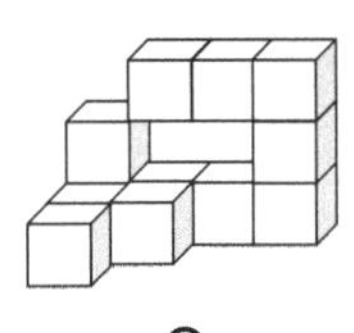
Q

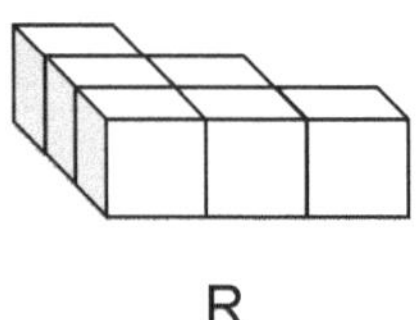
R

A O + P= Q + R + 1
B O + P = Q + R - 1
C P + Q = R + O
D P + Q = R + O + 5
E none of the above

33 The area of the shaded part of this square is 33 cm^2.

What is the area of the whole square?

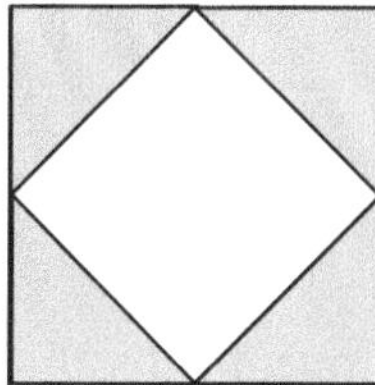

A 32 cm^2
B 48 cm^2
C 52 cm^2
D 66 cm^2
E 132 cm^2

34 James spent one-fifth of his money on a T-shirt and half the remainder on a pair of shoes.

If he has $40 remaining, how much did he start with?

A $100
B $120
C $150
D $200
E $240

35 The country of Wakanda has a currency called 'Wakam'.

They have coins of the following types:

- 1 wakam
- 2 wakams
- 5 wakams
- 10 wakams
- 20 wakams
- 50 wakams

Adam bought an item that cost 79 wakams and no change was required.

What is the smallest number of coins Adam could have used to pay?

A 7

B 6

C 5

D 4

E 3

OC Practice Test
Mathematical Reasoning 6 – Answer Sheet

Fill in the appropriate circle for your chosen answer

Eg.. A B C D E
○ ● ○ ○ ○

Use a pencil. If you make a mistake, erase thoroughly and try again.

NAME : **SCORE:**

1	A B C D E ○ ○ ○ ○ ○	8	A B C D E ○ ○ ○ ○ ○	15	A B C D E ○ ○ ○ ○ ○	22	A B C D E ○ ○ ○ ○ ○	29	A B C D E ○ ○ ○ ○ ○
2	A B C D E ○ ○ ○ ○ ○	9	A B C D E ○ ○ ○ ○ ○	16	A B C D E ○ ○ ○ ○ ○	23	A B C D E ○ ○ ○ ○ ○	30	A B C D E ○ ○ ○ ○ ○
3	A B C D E ○ ○ ○ ○ ○	10	A B C D E ○ ○ ○ ○ ○	17	A B C D E ○ ○ ○ ○ ○	24	A B C D E ○ ○ ○ ○ ○	31	A B C D E ○ ○ ○ ○ ○
4	A B C D E ○ ○ ○ ○ ○	11	A B C D E ○ ○ ○ ○ ○	18	A B C D E ○ ○ ○ ○ ○	25	A B C D E ○ ○ ○ ○ ○	32	A B C D E ○ ○ ○ ○ ○
5	A B C D E ○ ○ ○ ○ ○	12	A B C D E ○ ○ ○ ○ ○	19	A B C D E ○ ○ ○ ○ ○	26	A B C D E ○ ○ ○ ○ ○	33	A B C D E ○ ○ ○ ○ ○
6	A B C D E ○ ○ ○ ○ ○	13	A B C D E ○ ○ ○ ○ ○	20	A B C D E ○ ○ ○ ○ ○	27	A B C D E ○ ○ ○ ○ ○	34	A B C D E ○ ○ ○ ○ ○
7	A B C D E ○ ○ ○ ○ ○	14	A B C D E ○ ○ ○ ○ ○	21	A B C D E ○ ○ ○ ○ ○	28	A B C D E ○ ○ ○ ○ ○	35	A B C D E ○ ○ ○ ○ ○

BLANK PAGE

OC Practice Test
Mathematical Reasoning 7 – Answer Sheet

Fill in the appropriate circle for your chosen answer

Eg.. A B C D E
○ ● ○ ○ ○

Use a pencil. If you make a mistake, erase thoroughly and try again.

NAME : **SCORE:**

1	A B C D E ○ ○ ○ ○ ○	8	A B C D E ○ ○ ○ ○ ○	15	A B C D E ○ ○ ○ ○ ○	22	A B C D E ○ ○ ○ ○ ○	29	A B C D E ○ ○ ○ ○ ○
2	A B C D E ○ ○ ○ ○ ○	9	A B C D E ○ ○ ○ ○ ○	16	A B C D E ○ ○ ○ ○ ○	23	A B C D E ○ ○ ○ ○ ○	30	A B C D E ○ ○ ○ ○ ○
3	A B C D E ○ ○ ○ ○ ○	10	A B C D E ○ ○ ○ ○ ○	17	A B C D E ○ ○ ○ ○ ○	24	A B C D E ○ ○ ○ ○ ○	31	A B C D E ○ ○ ○ ○ ○
4	A B C D E ○ ○ ○ ○ ○	11	A B C D E ○ ○ ○ ○ ○	18	A B C D E ○ ○ ○ ○ ○	25	A B C D E ○ ○ ○ ○ ○	32	A B C D E ○ ○ ○ ○ ○
5	A B C D E ○ ○ ○ ○ ○	12	A B C D E ○ ○ ○ ○ ○	19	A B C D E ○ ○ ○ ○ ○	26	A B C D E ○ ○ ○ ○ ○	33	A B C D E ○ ○ ○ ○ ○
6	A B C D E ○ ○ ○ ○ ○	13	A B C D E ○ ○ ○ ○ ○	20	A B C D E ○ ○ ○ ○ ○	27	A B C D E ○ ○ ○ ○ ○	34	A B C D E ○ ○ ○ ○ ○
7	A B C D E ○ ○ ○ ○ ○	14	A B C D E ○ ○ ○ ○ ○	21	A B C D E ○ ○ ○ ○ ○	28	A B C D E ○ ○ ○ ○ ○	35	A B C D E ○ ○ ○ ○ ○

BLANK PAGE

OC Practice Test
Mathematical Reasoning 8 – Answer Sheet

Fill in the appropriate circle for your chosen answer

Eg.. A B C D E
○ ● ○ ○ ○

Use a pencil. If you make a mistake, erase thoroughly and try again.

NAME : **SCORE:**

1	A B C D E ○○○○○	8	A B C D E ○○○○○	15	A B C D E ○○○○○	22	A B C D E ○○○○○	29	A B C D E ○○○○○
2	A B C D E ○○○○○	9	A B C D E ○○○○○	16	A B C D E ○○○○○	23	A B C D E ○○○○○	30	A B C D E ○○○○○
3	A B C D E ○○○○○	10	A B C D E ○○○○○	17	A B C D E ○○○○○	24	A B C D E ○○○○○	31	A B C D E ○○○○○
4	A B C D E ○○○○○	11	A B C D E ○○○○○	18	A B C D E ○○○○○	25	A B C D E ○○○○○	32	A B C D E ○○○○○
5	A B C D E ○○○○○	12	A B C D E ○○○○○	19	A B C D E ○○○○○	26	A B C D E ○○○○○	33	A B C D E ○○○○○
6	A B C D E ○○○○○	13	A B C D E ○○○○○	20	A B C D E ○○○○○	27	A B C D E ○○○○○	34	A B C D E ○○○○○
7	A B C D E ○○○○○	14	A B C D E ○○○○○	21	A B C D E ○○○○○	28	A B C D E ○○○○○	35	A B C D E ○○○○○

BLANK PAGE

OC Practice Test
Mathematical Reasoning 9 – Answer Sheet

Fill in the appropriate circle for your chosen answer

Eg.. A B C D E
○ ● ○ ○ ○

Use a pencil. If you make a mistake, erase thoroughly and try again.

NAME : **SCORE:**

1	A B C D E ○ ○ ○ ○ ○	8	A B C D E ○ ○ ○ ○ ○	15	A B C D E ○ ○ ○ ○ ○	22	A B C D E ○ ○ ○ ○ ○	29	A B C D E ○ ○ ○ ○ ○
2	A B C D E ○ ○ ○ ○ ○	9	A B C D E ○ ○ ○ ○ ○	16	A B C D E ○ ○ ○ ○ ○	23	A B C D E ○ ○ ○ ○ ○	30	A B C D E ○ ○ ○ ○ ○
3	A B C D E ○ ○ ○ ○ ○	10	A B C D E ○ ○ ○ ○ ○	17	A B C D E ○ ○ ○ ○ ○	24	A B C D E ○ ○ ○ ○ ○	31	A B C D E ○ ○ ○ ○ ○
4	A B C D E ○ ○ ○ ○ ○	11	A B C D E ○ ○ ○ ○ ○	18	A B C D E ○ ○ ○ ○ ○	25	A B C D E ○ ○ ○ ○ ○	32	A B C D E ○ ○ ○ ○ ○
5	A B C D E ○ ○ ○ ○ ○	12	A B C D E ○ ○ ○ ○ ○	19	A B C D E ○ ○ ○ ○ ○	26	A B C D E ○ ○ ○ ○ ○	33	A B C D E ○ ○ ○ ○ ○
6	A B C D E ○ ○ ○ ○ ○	13	A B C D E ○ ○ ○ ○ ○	20	A B C D E ○ ○ ○ ○ ○	27	A B C D E ○ ○ ○ ○ ○	34	A B C D E ○ ○ ○ ○ ○
7	A B C D E ○ ○ ○ ○ ○	14	A B C D E ○ ○ ○ ○ ○	21	A B C D E ○ ○ ○ ○ ○	28	A B C D E ○ ○ ○ ○ ○	35	A B C D E ○ ○ ○ ○ ○

BLANK PAGE

OC Practice Test
Mathematical Reasoning 10 – Answer Sheet

Fill in the appropriate circle for your chosen answer

Eg.. A B C D E
○ ● ○ ○ ○

Use a pencil. If you make a mistake, erase thoroughly and try again.

NAME : **SCORE:**

1 A B C D E ○○○○○	8 A B C D E ○○○○○	15 A B C D E ○○○○○	22 A B C D E ○○○○○	29 A B C D E ○○○○○
2 A B C D E ○○○○○	9 A B C D E ○○○○○	16 A B C D E ○○○○○	23 A B C D E ○○○○○	30 A B C D E ○○○○○
3 A B C D E ○○○○○	10 A B C D E ○○○○○	17 A B C D E ○○○○○	24 A B C D E ○○○○○	31 A B C D E ○○○○○
4 A B C D E ○○○○○	11 A B C D E ○○○○○	18 A B C D E ○○○○○	25 A B C D E ○○○○○	32 A B C D E ○○○○○
5 A B C D E ○○○○○	12 A B C D E ○○○○○	19 A B C D E ○○○○○	26 A B C D E ○○○○○	33 A B C D E ○○○○○
6 A B C D E ○○○○○	13 A B C D E ○○○○○	20 A B C D E ○○○○○	27 A B C D E ○○○○○	34 A B C D E ○○○○○
7 A B C D E ○○○○○	14 A B C D E ○○○○○	21 A B C D E ○○○○○	28 A B C D E ○○○○○	35 A B C D E ○○○○○

BLANK PAGE

OC PRACTICE TEST ANSWERS

OC Practice Test Answers

Mathematical Reasoning 6

Question	Answer	Explanation
1	A	Factors of 12: 1,2,3,4,6,12 Factors of 15: 1,3,5,15 Factors of 16: 1,2,4,8,16 Factors of 19: 1,19 Factors of 21: 1,3,7,21
2	B	
3	C	Add one then square the number.
4	E	10 out of 16 = 5/8
5	B	A(1) B(2) C(3) D(4) E(5) F(6) G(7) H(8) I(9) J(10) K(11) L(12) M(13) N(14) O(15) P(16) Q(17) R(18) S(19) T(20) U(21) V(22) W(23) X(24) Y(25) Z(26)
6	B	a circle: 5 a triangle: 7 a black shape: 6 a white shape: 6 a white triangle: 4
7	E	Jason's height increased by 40 cm $\frac{1}{3}$ of 120 cm = 40 cm
8	C	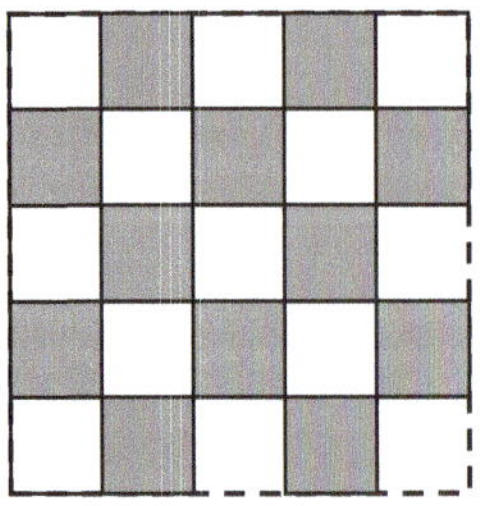
9	A	3 × \$45 000 = \$135 000
10	C	135 000 + 120 000 + 80 000 + 100 000 = 435 000

<table>
<tr><td>11</td><td>B</td><td>He has $65 000 left over. The most he can afford is a bathroom area.</td></tr>
<tr><td>12</td><td>C</td><td>(6 x 4) + (8 x 9) = $96m^2$</td></tr>
<tr><td>13</td><td>D</td><td>6 crates - He needs 5 crates for water, and 1 crate for the soft drink.</td></tr>
<tr><td>14</td><td>B</td><td>Every increment is 0.2 units, therefore the arrow indicates 12.9 units.</td></tr>
<tr><td>15</td><td>B</td><td>Total time needed = 27 + 14 = 41 minutes
Must leave no later than 4:19
Therefore, the ideal train would be 4:13</td></tr>
<tr><td>16</td><td>E</td><td></td></tr>
<tr><td>17</td><td>C</td><td>Since she started on Friday, there will be 5 Sundays among 31 days.
Therefore, (5 x 5) + (3 x 26) = 103</td></tr>
<tr><td>18</td><td>B</td><td>Test it. Say our three digit number is 100. If we add 4 to the back we get 1004.
100 x 10 + 4 = 1004.</td></tr>
<tr><td>19</td><td>B</td><td><table>
<tr><td>Black</td><td>1</td><td>2</td><td>3</td><td>4</td><td>5</td><td>6</td><td>7</td><td>8</td><td>9</td><td>10</td><td>11</td></tr>
<tr><td>White</td><td>5</td><td>7</td><td>9</td><td>11</td><td>13</td><td>15</td><td>17</td><td>19</td><td>21</td><td>23</td><td>25</td></tr>
</table></td></tr>
<tr><td>20</td><td>B</td><td>There are 2 girls to the left, thus there are 2 boys also.
There are a total of (13 - 1) ÷ 2 = 6 boys, therefore there are 4 boys to her right.</td></tr>
<tr><td>21</td><td>C</td><td>The squares that are needed are: second row third from right, third row from bottom third column from right, fourth column from left bottom row.</td></tr>
<tr><td>22</td><td>E</td><td>Add former two numbers</td></tr>
<tr><td>23</td><td>A</td><td>TOTAL: 4 pens 6 erasers 5 pencils
R has 4 erasers, so N has 2 erasers and 0 pencils. J thus has 2 pencils and 0 erasers and 0 pens. R must then have 3 pencils and 3 pens. Therefore, N has 1 pen.</td></tr>
<tr><td>24</td><td>B</td><td>Use multiplication 128 x 7 = 896</td></tr>
<tr><td>25</td><td>D</td><td>Chloe needs the last 20%. 0.2 x $700 = $140 still to be saved.</td></tr>
</table>

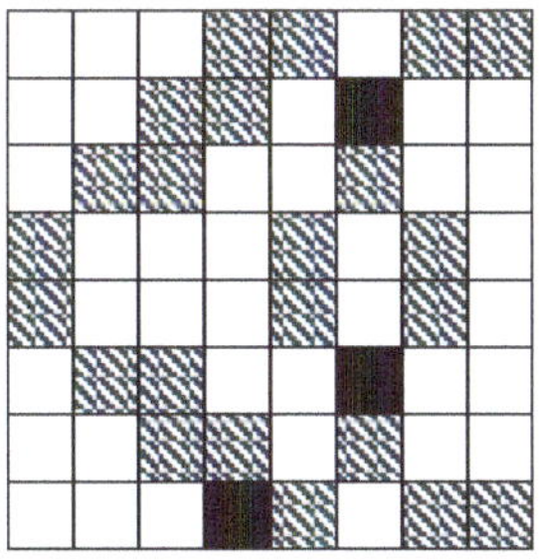

26	C	21/3 * 5 = 7 *5 =35 6 * 5 = 30 35>30
27	C	If a quarter is underground, then 3/4 is above the ground. Therefore 3.6m x 4/3 = 4.8m
28	A	Halve the sum of the numbers in the first two boxes to get the number in the last box.
29	D	40704, 41514, 42324, 43134
30	A	1st column: pentagon + 2 circle = 18 2nd row: 2 pentagon + circle = 21 Since pentagon = 18 – 2 circle Then, 36 – 4 circle + circle = 21 3 circle = 15 Circle = 5
31	D	4 strokes = 6m. 1200 strokes = 4 strokes x 300 = 6m x 300 = 1800m. Therefore, Dustin rowed 1.8km.
32	C	2 x 23 + 4 = 50 2500/50 = 50
33	B	Ben $(\frac{1}{6})$ + Nina $(\frac{2}{6})$ = $\frac{3}{6}$ = $\frac{1}{2}$ Ali $(\frac{1}{12})$ + Sarah $(\frac{5}{12})$ = $\frac{6}{12}$ = $\frac{1}{2}$ Sarah $(\frac{5}{12})$ => 10 berries Ali $(\frac{1}{12})$ => 2 berries Ben $(\frac{2}{12})$ => 4 berries Nina $(\frac{4}{12})$ => 8 berries 10 + 2 + 4 + 8 = 24
34	C	For the average of 10 students to be 72, total must be 10x72 = 720 Currently class of 9 students' average is 70, total is 9x70 = 630 10th student needs mark of 720 – 630 = 90
35	B	6,7,9,10,11,12,13,15

OC Practice Test Answers

Mathematical Reasoning 7

Question	Answer	Explanation
1	D	7 x 3 = 21
2	C	
3	A	Add consecutive square numbers: +1, +4, +9, +16, +25, +36
4	B	! + ! = 18 ! = 9 ? = 3 * + * + 9 = 21 * + * =12 * = 6
5	A	6 x 5 - 7
6	B	We can find the length of the missing at the bottom because 10 + ? = 4 + 5 + 5. Therefore, the missing side is 4. We can do the same to find the missing side on the right because 8 + ? = 8 + 3 + 3. Therefore, the missing side is 6. Now that we have all the sides, we can add it all up to find the perimeter: 14 + 14 + 14 + 14 = 56
7	A	D1+(D1+10)+(D1+20)+(D1+30)+(D1+40) = 125 5 x D1 = 25 therefore D1 = 5 Therefore, D3 = 5 + 20 = 25
8	E	863 + 1 = 864
9	B	4422 / 11 = 402
10	A	360/(3x60) = 2
11	B	45678 x 0.01 = 456.78
12	A	1 - 3/7 = 4/7 210 * 4/7 = 120 1/3 * 120 = 40
13	C	20 – 40 (6), over 40 (5)

14	E	Dennis has 60 cards which he sells at 6 for \$3.50. To find out how much money he makes from selling them, multiply \$3.50 by 10 = \$35. Subtract \$24 from \$35 to find his profit = \$11
15	D	S – 6s – Ar – 2s – N – 8 s – J – 5 s – Am Therefore, Nikhil was 6 + 2 = 8 seconds slower than Sumal.
16	E	Daily 10pm – 6am (8 hrs) 8 hrs x 7 days = 56 hrs 56 hrs x 60 = 3360 minutes
17	C	Jackie can make 4pies in 6 minutes Lenny can make 9pies in 6 minutes They can make 13 pies in 6 minutes together Therefore, they can make 130 pies in 60 minutes (1hr)
18	A	7000 + 300 + 40 + 0.2 + 0.05 = 7340.25
19	C	The order from first to last: S, T, D, C, A, J. Therefore, Cate came fourth.
20	B	The minute hand moves a quarter of the way around the clock (which is a circle), thus EXACTLY 90 degrees.
21	B	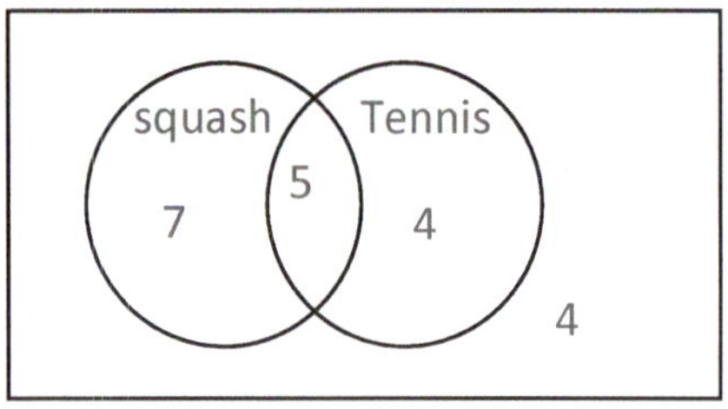 squash: 7; squash and Tennis: 5; Tennis: 4; neither: 4
22	C	Today is Saturday since tomorrow will be Sunday and 2 days before that is Friday (as required by the question). 8 days from Saturday is Sunday.
23	B	2 5 9 14 20 +3 +4 +5 +6
24	E	Let each term in brackets be one of the consecutive even numbers (x), (x+1), (x+2)... (x)+ (x+1)+(x+2) = 15 3x + 3 =15 3x = 12 X=4 (the first number of the MIDDLE 3 numbers of the set of 5.) Therefore, the numbers are 3, 4, 5, 6, 7. And the 4th number is 6.
25	A	5 blocks in Q, and 3 × 4 × 5 blocks in T

<table>
<tr><td>26</td><td>C</td><td>A + 6A + 3A = 30
A = 3
B = 18
A + B = 21</td></tr>
<tr><td>27</td><td>D</td><td>20 chocolate chips for 100g of cookie mixture
1 chocolate chip for 5g of cookie mixture
152 chocolate chips for 760g of cookie mixture</td></tr>
<tr><td>28</td><td>A</td><td>The 9 in 7681.29 is in the second decimal place. Therefore, its value is 9x 0.01 because 0.01 is also the second decimal place.</td></tr>
<tr><td>29</td><td>D</td><td>First 4 even numbers and zero = 2, 4, 6, 8, 0
Largest = 86420
Second = 86402
Third = 86240</td></tr>
<tr><td>30</td><td>A</td><td>Separate the figure into tenths by drawing a line horizontally across the middle and see how much of it is shaded.

4 out of 10 squares are shaded = 4/10 = 2/5</td></tr>
<tr><td>31</td><td>E</td><td>2B + 1D = $10.50
+ 1B + 2D = $7.50
3B + 3D = $18
1B + 1D = $6</td></tr>
<tr><td>32</td><td>C</td><td>Consecutive order of prime numbers. The one after 7 is 11.</td></tr>
<tr><td>33</td><td>B</td><td>7+7+5+5 = 24</td></tr>
<tr><td>34</td><td>C</td><td>250 + 400 + 750 + 400 + 400 + 400 + 400 = 3000g</td></tr>
<tr><td>35</td><td>E</td><td>Alan needs to save another 40% of $600</td></tr>
</table>

OC Practice Test Answers

Mathematical Reasoning 8

Question	Answer	Explanation
1	D	
2	C	Last two 2d shapes do not have right angle.
3	B	Max 62kg, Justin 32kg, Jim 17kg 62-17 = 45kg
4	C	LCM of 3,4,5 is 60
5	E	80 m x 5 = 400 m
6	B	**4199** is the closest number to 4200 without being 4200 or over ("**almost** double were sold")
7	A	8 sets of 2 blocks. 8x5=40
8	D	North - > East -> West
9	D	5 x 3 = 15
10	D	7 x 3 = 21
11	E	W = $\frac{1}{2}$, X = $\frac{3}{4}$, Y = $\frac{6}{10}$ = $\frac{3}{5}$, Z = $\frac{9}{12}$ = $\frac{3}{4}$
12	B	8 + 1 + 14 = 23
13	B	(6 x 25) – 90 = 60 cm (used it for knot) 60 cm ÷ 5 = 12 cm
14	A	2A + 1Or = \$3.00 + 1A + 2Or = \$3.30 3A + 3Or = \$6.30 1A + 1Or = \$2.10 Therefore, 1A = \$0.90
15	C	5.5hr x 60 = 330 minutes
16	A	4 × 6 × 2 = 48
17	A	3/2 x 12 = 18

18	D	1 circle = 5 triangles Therefore, 3 circles = 15 triangles
19	B	37 – 13 = 24 24 ÷ 6 = 4 (each time increases by 4) Therefore, 37(9th), 41(10th), 45 (11th), 49 (12th)
20	D	Least number of students take train.
21	D	$20 + 20 \times \frac{1}{2} + (20 \times \frac{1}{2} \times \frac{1}{2}) = 20 + 10 + 5 = 35$ seconds.
22	E	The volume of the driveway is 12 × 15 × 0.2 = 36m^3, the amount of sand is one quarter of that which is 9 m^3, equalling 9 000 000 cm^3.
23	A	The pattern is: doubling the numbers each time. The next number will be 64 x 2 = 128
24	E	X $\frac{2}{5} + \frac{2}{5} = \frac{4}{5} > \frac{3}{5}$ Y $1 - \frac{1}{3} = \frac{2}{3} > \frac{1}{3}$ Z $\frac{2}{4} \times \frac{1}{2} = \frac{1}{4} > \frac{1}{8}$
25	C	4 m per 10 seconds = 400 cm per 10 seconds = 40 cm per second
26	A	33.3 – 3.33 = 29.97
27	D	Triangle: 2cm x 2cm ÷ 2 = $2cm^2$ Rectangle: 3cm x 2cm = $6cm^2$ Therefore, $2cm^2 + 6cm^2 = 8cm^2$
28	B	3948/7 = 564. Mean of digits is (5+6+4)/3=**5**
29	D	Original number of bikes = 420 - 220 = 200 Original number of cars = 200 + 50 = 250 Current number of cars = 250 – 100 = **150**
30	C	12.4-1.3-3.2-3.6=4.3
31	B	(16 x 5) – (4 x 4) = 64
32	E	V = 4 x 3 x 4 Y = 3 x 4 x 4

33	B	X $\frac{1}{5}$ = 20% Y $\frac{4}{5}$ = 80% Z $\frac{0}{5}$ = 0%
34	A	6 + 8 + 13 + 9 = 36
35	E	$\frac{3}{5}$ kg of meat => 4 hamburgers To make 80 hamburgers, $\frac{3}{5}$ kg x 20 = 12kg

OC Practice Test Answers

Mathematical Reasoning 9

Question	Answer	Explanation
1	D	6 x 2 x 15 = 180
2	D	4 + 8 x 3 = 28 = 37 - 9
3	E	(L x 6) + (W x 4) = 80 cm L = 2W 12W + 4W = 80 cm 16W = 80 cm W = 5 cm L = 10 cm
4	A	$\frac{6}{8}$ of x = $\frac{3}{4}$ of 32 = 24 X = 32
5	B	B is the only solution which fits the information given.
6	B	Doll (85g) + car (46g) + ball (65g) = 196g
7	A	Add all the numbers on the outside. Then divide by two to get the number in the middle of the triangle. (16+8+?)/2 = 14 16 + 8 + ? = 28 Therefore, ? = **4**
8	B	(5-2) + (4-1) = **6m** in **total**
9	B	2 x 1 = 2 ways
10	C	First, third and fifth 2D shapes are parallelograms.
11	A	J-1=(D-1)/3 40=D+3 D=37 J-1=12 J+3=16
12	C	Area of triangle = 2cm x 2cm ÷ 2 = $2cm^2$ Area of rectangle = 4cm x 8cm = $32cm^2$ 32 ÷ 2 = 16
13	E	30 – 5 = 25 (15 + 15) – 25 = 5

14	B	It will become 3 by 3 by 4 rectangular prism (3x3x4)-2=34
15	B	One box weighs 7kg
16	E	Count it
17	D	Krishni gets 20, Emily gets 5
18	B	80% of 3 tests require 96 questions correct. Therefore, 96-30-28 =38
19	C	[2]O[10] [12]O[12] 10 more
20	A	20L Hence 8 buckets of 2.5L
21	C	28 ÷ 3 = 7r1 28 ÷ 8 = 3r4
22	E	South-west -> north-west
23	A	7 parts, divided into 3 parts each
24	C	3,6,9,12,15
25	C	For the average of three numbers to be 30, they must add up to 90. The numbers add up to 90 in C. Alternatively, since 30 is the average you want, in B you have one number above and one number below 30: 31 and 29. This averages out to give you three 30s.
26	B	The left hand side is the same as 11 diamonds. One diamond is therefore 11 kg.
27	A	A = 20cm G = 80cm N = 60cm N = 15 x 0.6m = **9**m
28	D	A, B, C cannot be proven as comparatively there is no trend. **D** is the most correct as comparatively, Jane is taller than everyone else although she is the youngest.
29	C	3.2m – (0.5cm x 4) = 1.2m
30	E	40% of 50 students = 20
31	C	21 x 5 = 105
32	C	10 x 5 = 50

33	D	$\frac{2}{5}$ of total points now since he needs another $\frac{3}{5}$ of total points until levelling up. $\frac{2}{5}$ of total points = 20 Total points = 50
34	C	(\$8.40 x 5) + \$12.40 + 2 novels = \$75 \$42 + 12.40 + 2 novels = \$75 \$54.40 + 2 novels = \$75 2 novels = \$20.60 1 novel = \$10.30
35	E	1 5 students (3 siblings), 2 students (2 siblings) 2 10 students (2 siblings), 5 students (3 siblings) 3 15 students (1 sibling), 2 students (4 siblings)

OC Practice Test Answers

Mathematical Reasoning 10

Question	Answer	Explanation
1	B	12 x 2 = 24
2	A	The two lines in A form right angles with each other.
3	D	5 canaries out of all 9 birds
4	E	Shape 1 = 5, Shape 2 = 1, Shape 3 = 3, Shape 4 = 0, Shape 5 = 1
5	B	B ran the second furthest with the least amount of water.
6	B	(2x28) – 38 = 18
7	C	The 3D P has 12 faces.
8	B	$4 \times 7 - \frac{2}{3} \times 9 = 28 - 6 = 22$
9	B	1 Mary = $\frac{1}{2}$ 2 Jessica = $\frac{3}{8} > \frac{1}{4}$ (only true) 3 Vivian + Jessica = $\frac{1}{2}$
10	D	The time period is 40 ÷ 8 = 5 times the original length. In this time Joshua serves 5 × 6 = 30 customers.
11	B	1.12 + 0.88 = 2
12	E	4(small) + 3(middle) + 2(big) +1(biggest)
13	D	30 / 5 =6, Jess will receive 6 free balls 30 + 6 = 36
14	C	(8 x 8) – (4 x 4) = 48cm^2
15	B	45/5 = 9 chocolates 45/3 = 15 marshmallows 15-9=6 Therefore, there were **45** chocolates to begin with.

16	D	$1 - \frac{5}{8} = \frac{3}{8}$ leftovers $\frac{3}{8}$ of whole cake = 600g Whole cake = 1600g = 1.6kg
17	C	C is the only answer which matches the question
18	A	3 boxes coincide.
19	A	The last coordinate is B8.
20	B	4km+2km =6km 4km 3km 2km start 5km
21	E	76.5 – 69.8 = 6.7kg
22	A	This sequence contains an alternating pattern. The first pattern starts with 2 and increases by 2. The second pattern starts with 5 and increases by 1. The number we are looking for belongs to the first patter: 10+2=**12**.
23	C	B and D are the only shapes that will fit together.
24	D	1992/6 = 332
25	A	Kathy and her elder sister received \$80 × $\frac{3}{4}$ = \$60. Kathy got one quarter of this, totalling \$15.
26	C	4 x 2 + 1 = 9, 9 x 2 + 1 = 19, 19 x 2 + 1 = 39
27	A	$3 \times \frac{7}{8} = \frac{21}{8} = 7 \times \frac{3}{8}$
28	D	B and D will both have 7/16 shaded.
29	C	Jake can make 40 pies in 60 mins Lenny can make 60 pies in 60 mins Therefore, they can make 100 pies in 60 mins together.
30	C	\$50-\$27=\$23

31	D	Total weight of 30 pineapples = 30 x 0.5kg = 15kg Joyce kept $\frac{1}{3}$ of 15kg = 5kg (10kg remaining) Michael bought $\frac{1}{2}$ of 10kg = 5kg
32	B	O = 7, P = 10, Q = 12, R = 6, O + P = Q + R − 1
33	D	The area of half of the square is 33 units, therefore the whole square will be 33 x 2 = 66 units squared.
34	A	He spent $\frac{1}{5}$ on a shirt and $\frac{2}{5}$ on his shoes. $\frac{2}{5}$ of his original amount is \$40, therefore he started with \$100.
35	C	50 wakam + 20 wakam + 5 wakam + 2 wakam + 2 wakam = 79 wakams